KERATINIZATION

A Survey of Vertebrate Epithelia

PAUL F. PARAKKAL *and* NANCY J. ALEXANDER

Oregon Regional Primate Center
Beaverton, Oregon

ACADEMIC PRESS New York and London

ACADEMIC PRESS, INC.
111 Fifth Avenue, New York, New York 10003

United Kingdom Edition published by
ACADEMIC PRESS, INC. (LONDON) LTD.
24/28 Oval Road, London NW1

Library of Congress Catalog Card Number: 72-80256

CONTENTS

PREFACE

The skin of living vertebrates culminates a long history of evolution and adaption for survival in a hostile or at least potentially threatening environment. Vertebrate skin is the first line of defense in a continuing struggle for survival in water, on land, and in the air. Terrestrial vertebrates, in particular, display a great variety of integumentary modifications; scales, claws, feathers, beaks, hooves, horns, burrs, nails, quills, hair, and wool attest to their ability to deal effectively with heat and cold, enemy and prey, air and water, and all manner of terrestrial surfaces.

In this monograph, we have tried to illustrate the salient features of the epithelial portion of the skin and its appendages in each class of vertebrates. The emphasis has been ultrastructural, but we have tried to give the functional implications wherever possible, although these are often still conjectural.

In the Introduction, we emphasize the general process of keratinization; each subsequent section deals with aspects peculiar to a group of vertebrates. The monograph is designed to provide an introduction to the study of vertebrate skin and to stimulate professional investigators to delve into its mysteries, where mitosis, differentiation, and death occur sequentially and in close proximity.

Paul F. Parakkal
Nancy J. Alexander

ACKNOWLEDGMENTS

We are indebted to Mr. Joel Ito for the illustrations done with consummate patience and skill and to Dr. Ray Henrikson and Dr. Mary Bell for their micrographs of fish and fetal epidermis, respectively. We are likewise grateful to both Dr. W. H. Fahrenbach and Dr. William Montagna for their valuable suggestions and to Mrs. Margaret Barss for her editorial advice. Our efforts have been considerably lightened by the technical assistance of Mrs. Janice Anderson, Miss Nancy Trachsel, and Miss Jeanne Hren and by the secretarial services of Mrs. Sharon Maher. We thank the publishers who have allowed us to reproduce their figures.

I. INTRODUCTION

Vertebrate integument and its derivatives safeguard the animal from the many dangers of his environment. Because of the remarkable adaptations of skin, vertebrates have managed to survive and to thrive in water, on land, and in the air. Land animals in particular display an astonishing variety of integumentary modifications including scales, claws, feathers, beaks, hooves, horns, burrs, nails, quills, hair, and wool.

Skin and its appendages consist of two distinct components: surface epithelium, which develops from the embryonic ectoderm, and connective tissue, which develops from the mesoderm. These two components are mutually dependent for growth, differentiation, and function (2). In this monograph, epithelial differentiation and specialization in the various classes of vertebrates will be the primary concern.

The epithelium and its derivatives are built upon a common plan: all have a permanent population of germinal cells at the base whose progeny, as they migrate upward, undergo a specific pattern of differentiation (keratinization), at the end of which they are exfoliated (16). Thus, the life cycle of an epithelial cell consists of three phases: (a) mitosis, (b) differentiation, and (c) exfoliation (see Sections I,A,B,C).

A. Mitosis

The epithelium and its derivatives are self-renewing systems characterized by two patterns of growth. In the first, cells continuously undergo mitosis at the base of the epithelium and exfoliate at the surface. The various stratified epithelia of mammals and the epidermis of birds, amphibians, and fish are good examples of this pattern of growth. In the second pattern, periods of active growth with a burst of mitotic activity in the germinal population are followed by periods of quiescence. Squamate epidermis and hair follicles exemplify this growth pattern.

In the epithelium, the rate of mitosis delicately balances the loss of cells at the surface. However, such experimental procedures as wounding, radiation, or the removal of the superficial cells by stripping with tape or other methods can drastically increase mitotic rates. After the repair process, mitotic activity returns to normal. The rate of mitosis is not the same for all keratinizing epithelia. For example, renewal of the vaginal epithelium in the rodent occurs about every 4 to 6 days, that of the epidermis about every 20 to 25 days. Each tissue has a predetermined rate and any change causes a pathological condition.

The metabolic pathway of nucleic acids in the epidermis is the same as that of other tissues. During early development, labeled mitotic figures occur at all levels, but as the embryo grows they are progressively restricted to the base until adulthood when DNA replication (indicated by the incorporation of labeled thymidine) and mitotic figures are found almost exclusively in the basal layers.

The basal cells have an embryonic appearance with few distinctive intracellular differentiations and a nucleus that fills almost the entire cell. Ribosomes, the predominant cytoplasmic components of keratinizing cells, occur singly or in polysomal aggregates. The Golgi complex is usually small and located in a supranuclear position. Since keratinizing cells retain most differentiation products, they do not have the elaborate membrane system characteristic of cells that synthesize and secrete quantities of protein. Recent evidence, however, suggests that during development and wound repair basal cells secrete components that make up both the basal lamina (a 500 Å polysaccharide-rich layer under the epidermis) and the basal lamella. Under these conditions, the basal cells have many profiles of rough endoplasmic reticulum and a conspicuous Golgi complex (10).

B. Differentiation

As the epithelial cells become less mitotically active, specialization begins and they become involved in the synthesis of both fibrous and amorphous proteins; in addition, the surfaces of the cells become modified (2, 27). Finally, the nucleus and cytoplasmic organelles are gradually resorbed, partially or completely. This sequence has four phases: (a) fibrogenesis, (b) amorphous matrix formation, (c) cell surface modification, and (d) nuclear and organelle resorption.

1. Fibrogenesis

One of the main products of the cells of keratinizing tissue is fibrous protein. Early investigators, noting that the cornified layers of epidermis, hair, and wool are birefringent, concluded that the cells of horny tissues contain tonofibrils (*19*). These tissues also give strong histochemical reactions for S—S and S—H groups, an indication that the proteins are cross-linked through cystine sulfur. Astbury and co-workers, using x-ray diffraction techniques to elucidate the configuration of fibrous proteins in keratinizing tissues, found that hair, wool, and mammalian epidermis give an α-diffraction pattern similar to that of myosin and fibrinogen (*24*). Hence, keratin, myosin, epidermis, and fibrinogen are classified as K-M-E-F groups of proteins. When stretched, hair and wool, like the others in the K-M-E-F group give a β-diffraction pattern. This α to β transformation probably results from the extension of the folded polypeptide chains of the α-type proteins.

The x-ray diffraction patterns of the keratinizing tissues of vertebrates fall into three groups: (*a*) the α pattern (the epidermis of amphibians, birds, and mammals); (*b*) the β pattern (feathers, beaks, and turtle shell); (*c*) both patterns (the epidermis of lizards, snakes, and crocodilians) (*23*). Investigators have shown with the electron microscope that the cells of keratinizing tissues contain microfibrils whose frequencies of occurrence and modes of aggregation vary in different tissues. Tissues that produce α- and β- diffraction patterns contain microfibrils of 80 and 30 Å, respectively (*9, 12*). According to Filshie and Rogers (*7*), 80 Å microfibrils are composed of smaller units, 20 Å in diameter, called protofilaments, which form a circular configuration; however, the exact number and arrangement of these protofilaments remain controversial (*18*).

Based on the discovery by Pauling and Corey of the α-helical structure of protein chains, several models showing the number and configuration of polypeptide chains in natural keratins have been proposed. Pauling and Corey (*21*) suggested a seven-stranded model, Crick (*4*) proposed a two- or three-strand rope coiled upon itself. This coiled-coil model explains satisfactorily the 5.1 Å reflection anticipated from α-helical folding. How many of these coiled-coil strands would fit into the microfibril structure of keratinizing cells is still debatable.

As differentiation proceeds, microfibrils increase numerically until they fill the cell. The packaging of the microfibrils, however, varies in different keratinizing epithelia. In feathers, scales, cortex, and the stratum corneum of avian and mammalian epidermis, the cells are packed with microfibrils embedded in an amorphous matrix. In electron micrographs, these microfibrils look like electron-lucent areas surrounded by an electron-opaque matrix. The compact arrangement of this fibril–matrix complex, which is known as "keratin pattern" (*3*), is seen neither in the epidermis of amphibians and fish nor in the epithelium of mucous membranes, such as the vagina, esophagus, oral mucosa, and conjunctiva of mammals.

Attempts to isolate the biochemical components of microfibrils have been only partially successful, mainly because of the relative insolubility of the proteins of keratinizing tissues (*14*). However, a low sulfur substance, prekeratin, which can be drawn into fibers that give an α-type x-ray pattern, has been isolated from mammalian epidermis (*15*).

2. Amorphous Matrix Formation

Amorphous matrix proteins are a major component of keratinizing epithelia and are found in different proportions in the various keratinized cells (*5, 17*). In mammalian epidermis, hair, and wool cortex, the ratio of fibrillar to amorphous protein is about 1:1. On the other hand, the medulla and cuticle of hair are mostly amorphous proteins, with only traces of the fibrillar components. Biochemical studies have shown that the composition of amorphous protein is not uniform (*13*). The fraction (γ-keratose) obtained from wool and mammalian epidermis is high in sulfur, but medullary proteins have only traces of sulfur (*14*). Despite their toughness and resistance to most proteolytic agents, the amorphous proteins are a heterogeneous group.

Amorphous proteins are thought to consist of a wide variety of granules of different sizes and composition, including medullary and cuticular granules that fill the keratinized cells of the hair medulla and cuticle. In the stratified epithelia of many mammals, numerous amorphous keratohyalin granules fill the cells before complete cornification. Formed in the spinous layer, these granules grow into huge electron-dense masses. At one time, they were thought to be characteristic

Legend to Diagram

The central schematic diagram shows a keratinizing epithelium. The basal cells, resting upon a basal lamina, undergo mitosis (1). The cells above the basal layer begin to synthesize 80 Å filaments and matrix material (2). The filaments aggregate into bundles surrounded by the matrix (2). As differentiation continues, membrane-coating granules (2, A) are formed in association with the Golgi complex; subsequently, the granules migrate toward the apical cell surface and discharge into intercellular spaces. Concomitantly, small amorphous keratohyalin granules are formed, which increase in size and eventually occupy a large area of the cell (3, B). Once the differentiation products are formed, the nucleus and the organelles are resorbed (4). During the formation of the horny cells, the plasma membranes thicken (C). The fully cornified cells are filled with filaments surrounded by an amorphous matrix (5, D). The most superficial cells of the stratum corneum are exfoliated.

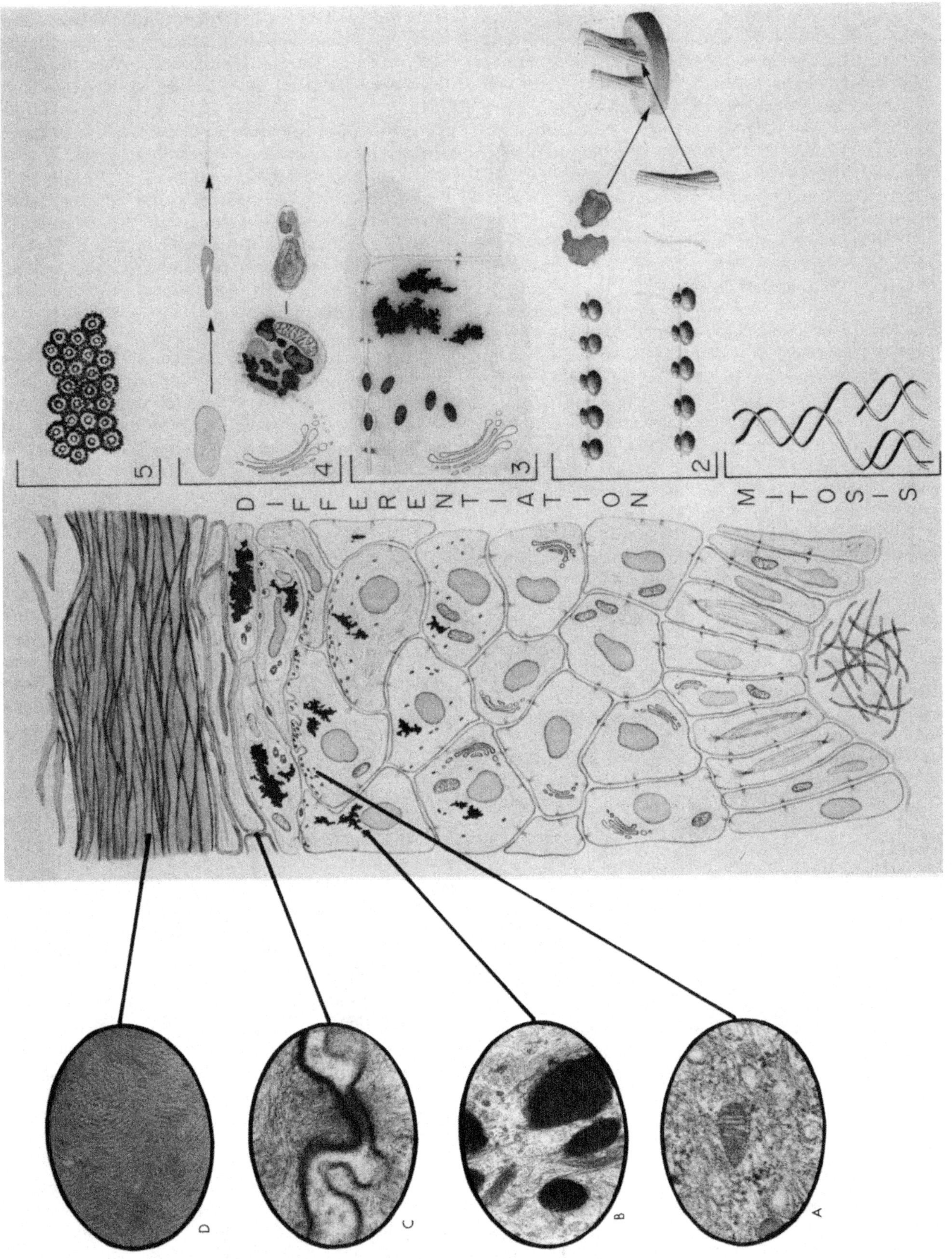
5
4
3
2
DIFFERENTIATION
MITOSIS
D
C
B
A

of mammalian epidermis, but recent electron microscope studies show them to be present in both avian and reptilian epidermis (1). Keratohyalin granules mysteriously disappear during the formation of cornified cells; some think that they transform into microfibrils, others that they become part of the matrix surrounding the microfibrils. Autoradiographic techniques have demonstrated that these granules contain a histidine-rich protein, whose role during keratinization remains an enigma.

Trichohyalin granules, identical morphologically with keratohyalin granules, appear in the cells of the inner root sheath during development. Like the keratohyalin granules, these granules also disappear as the cells of the inner root sheath become filled with microfibrils. Despite their morphological similarity, trichohyalin contains large amounts of arginine and is therefore chemically distinct from keratohyalin (22).

Electron microscopic studies have revealed a small component called membrane coating granules (MCG, Odland bodies, keratinosomes) in the differentiating cells of the mammalian keratinizing epithelia. These granules are ovoid in shape and measure 0.1 to 0.5 μ in diameter. Each granule is membrane bounded and closely packed with thick and thin lamellae which are about 30 Å. They are formed in close association with the Golgi complex and subsequently migrate toward the apical plasma membrane where they are discharged into the intercellular spaces. In the intercellular spaces the granules disintegrate and spread their contents over the plasma membranes. It is thought that these granules function as a cement binding the cells and acting as a water barrier.

Still another group of granules, known as multigranular bodies, have been shown to be present both in avian and reptilian epidermis but not in mammals. They consist of one to several lamellated bodies enclosed by a membrane and vary in size from 0.5 to 2 μ. These bodies begin to form in the layer above the basal cells and become incorporated into the horny cells.

In amphibians, mucous granules form at a level comparable with that of keratohyalin granules in the epidermis of birds and mammals and compose the interfilamentous amorphous matrix. As the horny cells form, these mucous granules disintegrate and disperse uniformly throughout the cytoplasm. A mucoprotein matrix is also found in certain mammalian stratified epithelia, such as fetal esophagus and certain parts of the conjunctiva.

3. Cell Membrane Modifications

As the epidermal cell moves toward the skin surface, the cell membranes change in thickness, resistance to keratolytic agents, and adhesive properties. The membranes of germinative and differentiating cells are about 80 Å thick and appear trilaminar and like those of other tissues. During keratinization, they double in thickness because of a thickening of the inner leaflet (6).

As the cell membranes thicken, they develop an extraordinary resistance to keratolytic agents. For example, high concentrations (0.1–0.4 N) of sodium hydroxide hydrolyze everything except cell membranes. Such resistant cell envelopes have been isolated from numerous tissues such as the epidermis of amphibians, reptiles, birds, and mammals and from hair and nail. Analysis shows isolated membranes to contain more sulfur than either the fibrous or amorphous components of horny cells.

In vertebrate epidermis, thickened cell membranes are common to all superficially placed keratinized cells; however, those of the β layer of reptilian epidermis do not retain recognizable cell membranes. During the final stages of maturation, the cell membranes lose their integrity and the whole layer looks like a syncytium.

4. Nuclear and Organelle Resorption

In keratinizing tissues like the epidermis of amphibia, reptiles, birds, and mammals and in hair and feathers, cytoplasmic organelles like mitochondria and ribosomes disappear when the cells are completely keratinized, leaving no trace. On the other hand, in the epidermis of fish and in the oral and esophageal epithelium of mammals, the fully developed cells retain many of the organelles. Evidence from several sources suggests that hydrolytic enzymes participate in the breakdown of the nucleus and cytoplasmic organelles. Histochemical studies show that the upper layers of stratified epithelia contain acid phosphatase, esterases, and β-glucuronidase (19, 20). Several investigators have reported autophagic vacuoles in the epidermis of amphibians and mammals (11). But the exact sequence of changes during the dissolution and resorption of the nucleus and cytoplasmic organelles is not known.

C. Exfoliation

At the end of differentiation, the cornified cells become the outer layer that protects the animal from the chemical and physical insults of the environment. The thickness of this layer varies in different vertebrates and in different parts of the same animal.

Like the stratified epithelia of mammals and the epidermis of birds and amphibians, the superficial cells of the horny layer are exfoliated continuously. However, both reptilian epidermis and mammalian hair are shed *en masse,* but not before new epidermis and new hair are formed.

How exfoliation of the superficial cells occurs is not fully understood, but recently it has been suggested that the membrane-coating granules (MCG, Odland bodies, keratinosomes) are involved in this process (*26, 28*). Histochemical studies have shown that the MCG's are positive for hydrolytic enzymes inside the cell and after being discharged into the intercellular spaces. A timely release of these enzymes in the upper stratum of the horny layer may be responsible for the orderly shedding of the horny cells. However, it is difficult to associate this function to membrane-coating granules in the formation of nail, where there is no exfoliation.

In conclusion, keratinization is a complex process involving first the synthesis of stabilized proteins that fill the cells to varying degrees to the exclusion of other cell constituents and, second, the formation of a highly resistant and modified plasma membrane. The result is the production of a tissue that is at once tough, insoluble, elastic, and eminently successful in affording protection to the animal.

In this introduction the general processes involved in keratinization have been emphasized. Naturally, in the different stratified epithelia of vertebrates, an enormous variation in keratinization is encountered. The body of the text is a comparative view of the similarities and differences in vertebrate epithelia with illustrative material exemplifying the salient features of the integument of each class of vertebrates. It would be impossible to illustrate all the varieties of each class; hence, the most representative or didactic samples of each class have been chosen.

II. FISH EPIDERMIS

All vertebrates are covered by a stratified epithelium, the main function of which is to protect the organism from its environment as well as to preserve the interior milieu. In fish, the superficial cells, which are in contact with an aqueous environment, do not keratinize as in reptiles, birds, and mammals. However, fish skin has also undergone adaptations, namely, the development of dermal scales and, in the epidermis, the formation of several cell types that perform special functions. The dermal scales are similar to bony structures and are not related to the epidermally derived scales of reptiles, birds, and mammals.

Fish epidermis is characterized by at least four cell types: keratinocytes (filament-containing cells), mucous cells, club cells, and chloride cells, the most prevalent of which is the keratinocyte (Fig. 1). All the basal cells are cuboidal and rest upon a basal lamina that is comparatively straight. Below the basal lamina, six to eight layers of orthogonally arranged collagen fibers form the basal lamella (Fig. 3).

Most of the organelles of the keratinocyte, the mitochondria and smooth endoplasmic reticulum, are crowded around the nucleus (Fig. 2). The rest of the cell is loosely packed with filaments about 80 Å in diameter. Even though the plasma membranes interdigitate extensively, there are only a few desmosomes.

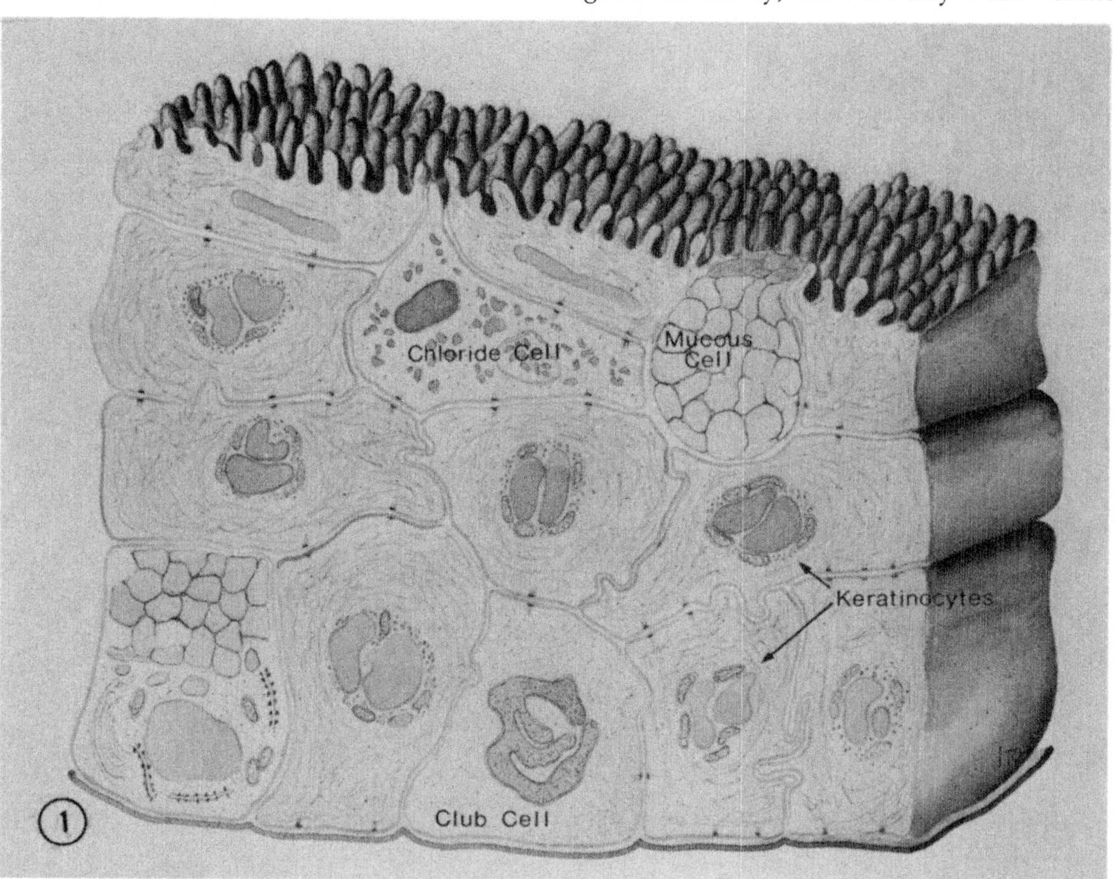

Fig. 1. Schematic diagram of fish epidermis showing keratinocytes and mucous, club, and chloride cells.

Fig. 2. The entire thickness of the epidermis of the guppy (*Lebistes reticulatus*) shows keratinocytes in various stages of differentiation. Veronal acetate buffered osmium fixation. ×1200. [From Henrikson, R. C., and Matoltsy, A. G. (1969). *J. Ultrastruct. Res.* **21,** 199. Courtesy of the authors and publisher.]

Fig. 3. Basal keratinocytes of the neon tetra (*Hyphessobrycon innesi*) resting on the orthogonally arranged collagen fibers of the basal lamella. Collidine buffered osmium fixation. ×25,000.

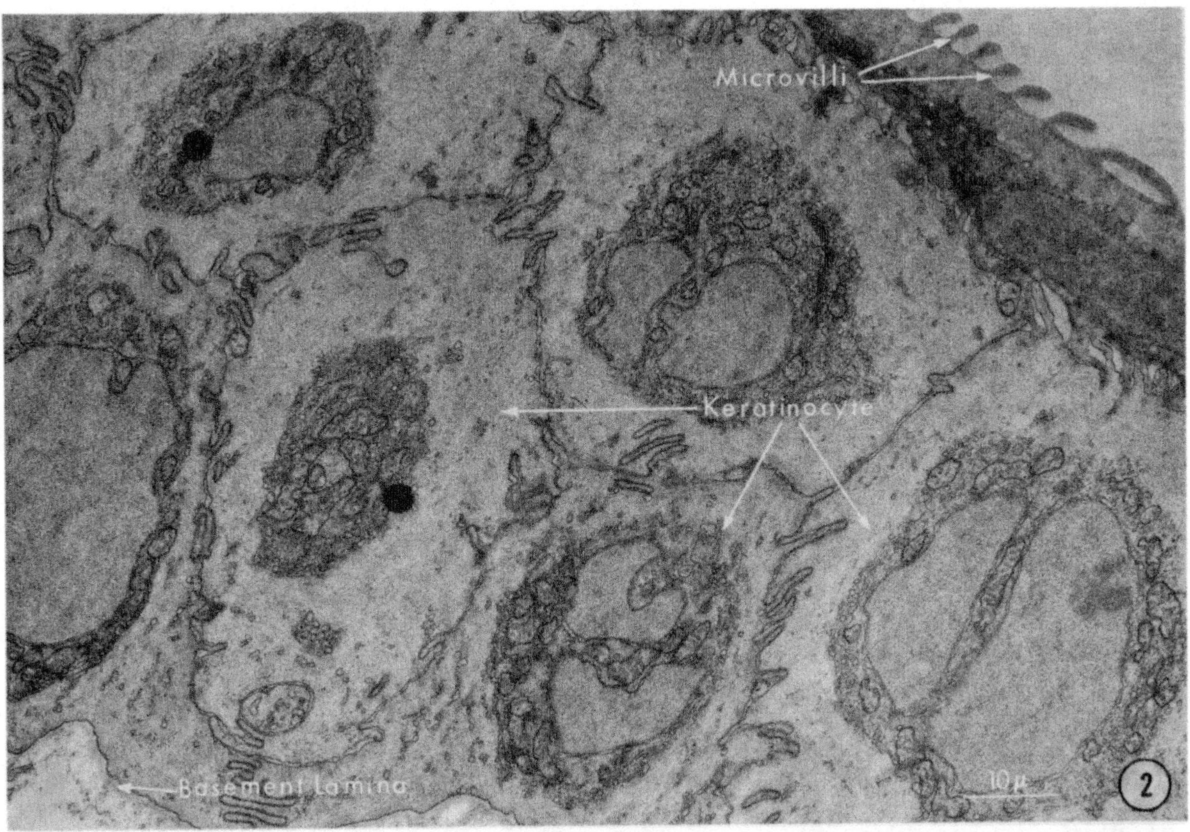

Microvilli
Keratinocyte
Basement Lamina
10 μ
2

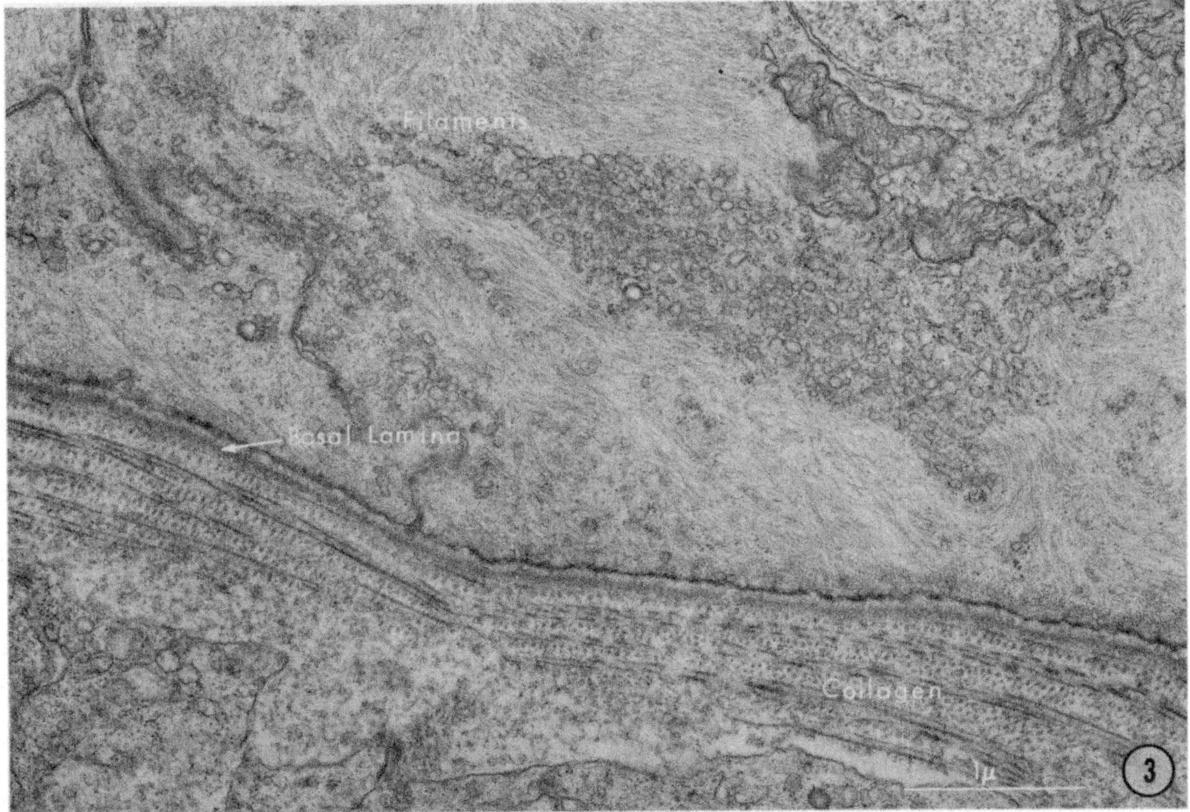

Filaments
Basal Lamina
Collagen
1 μ
3

These cells divide and as they migrate upward become the superficial cells. The apical border of these cells has an extensive microvillate surface; the number and height depend on the body location and the species of fish. The microvilli are covered with a layer of mucopolysaccharide filaments or fuzz that appear to be extensions of the outer leaflet of the plasma membrane (Fig. 4). The superficial cells are joined by junctional complexes that prevent seepage of fluid from the aqueous environment into the epidermal and the subjacent intercellular spaces. Unlike the superficial keratinized cells of most other vertebrates, the superficial cells of the fish epidermis are living and are even capable of incorporating tritiated thymidine, an indication of their potentiality to divide.

The most common unicellular gland of the fish epidermis is the mucous cell (Fig. 5). These cells originate basally and by the time they reach the surface are engorged with mucous droplets. The basal portion of the cell has an extensive rough-surfaced endoplasmic reticulum and Golgi complex. Mucous droplets develop in the cisternae of the endoplasmic reticulum and are elaborated further in the Golgi membranes, appearing like membrane-bound granules. In the fully developed cells, the mucous granules coalesce and their content is released over the surface of the fish. The secreted mucous certainly plays a protective role.

Fig. 4. Microvilli modifications and junctional complex of the superficial keratinocyte of the neon tetra (*Hyphessobrycon innesi*). Collidine buffered osmium fixation. ×12,000.

Fig. 5. Discharging mucous cell of the neon tetra (*Hyphessobrycon innesi*). Collidine buffered osmium fixation. ×12,000.

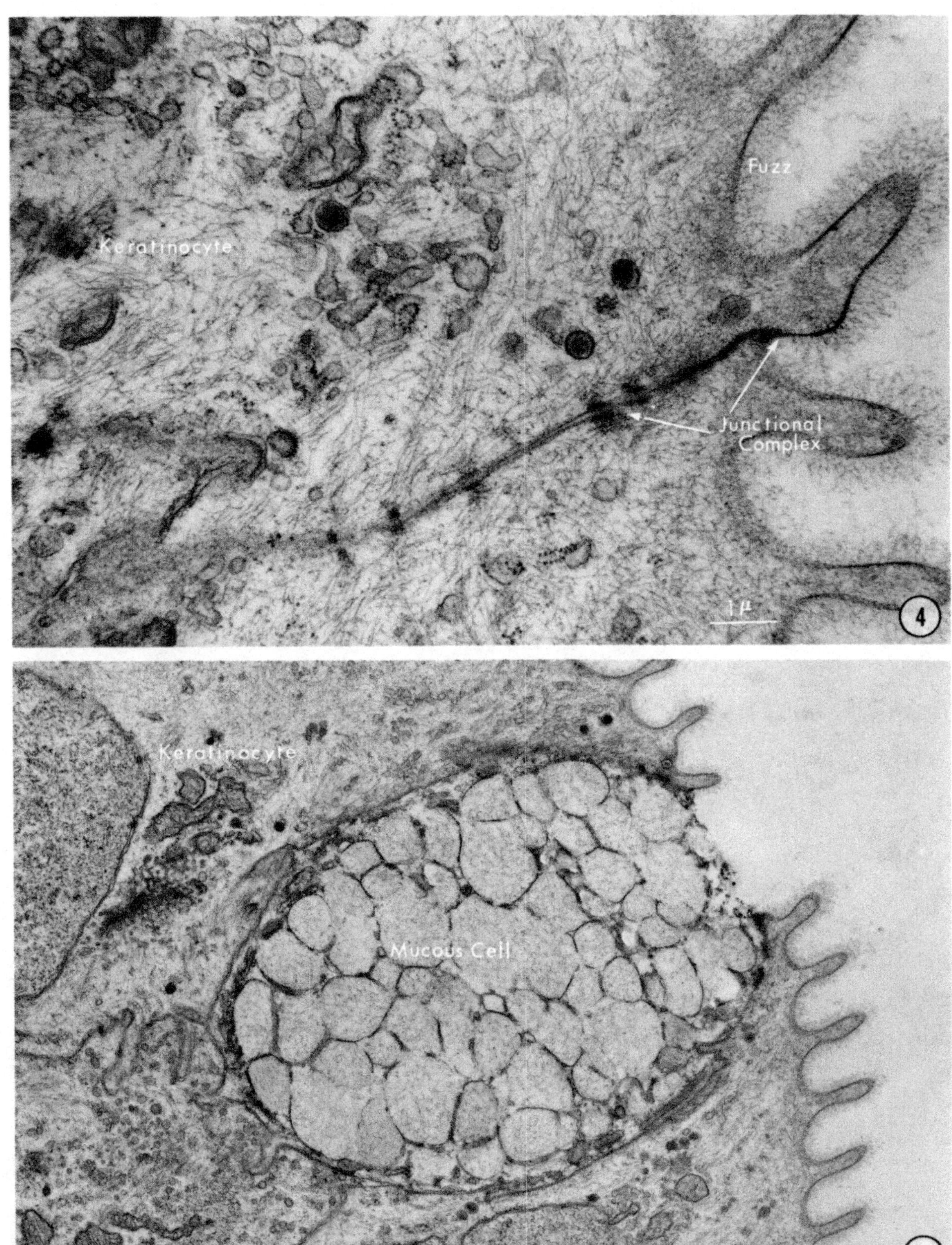

Fuzz
Keratinocyte
Junctional Complex
1μ
4
Keratinocyte
Mucous Cell
1μ
5

Among the unicellular glands in the teleost epidermis, club cells have multiple functions. In some teleosts, when threatened or wounded, the club cells release a pheromone-like substance into the surrounding water. This substance causes a fright reaction in other fish in the vicinity, exhibited by a sudden rapid swimming away from the location. In some eels, however, the secretion of the club cells forms part of the slime that covers the entire body of the fish. The most characteristic features of the club cells are their large size and irregularly shaped nucleus (Fig. 6). Most of the cytoplasm is occupied by a finely dispersed fibrillar material. The plasma membranes do not have the remarkable infoldings characteristic of keratinocytes.

Yet another type of cell common among marine fish is the chloride cell (Fig. 7). Although these cells are distributed over the whole body surface, they are most prevalent on the gills. Large amounts of smooth endoplasmic reticulum combined with numerous mitochondria distinguish this cell from the other cell types of fish epidermis. The main function of the chloride cell is to transport electrolytes during osmoregulation.

Fig. 6. Club cells of the catfish (*Corydoras aeneus*) showing the irregular nucleus and the finely dispersed filaments. Veronal acetate buffered osmium fixation. ×9000. (Courtesy of Dr. R. C. Henrikson.)

Fig. 7. Chloride cell flanked by keratinocytes in the epidermis of neon tetra (*Hyphessobrycon innesi*). Collidine buffered osmium fixation. ×24,600.

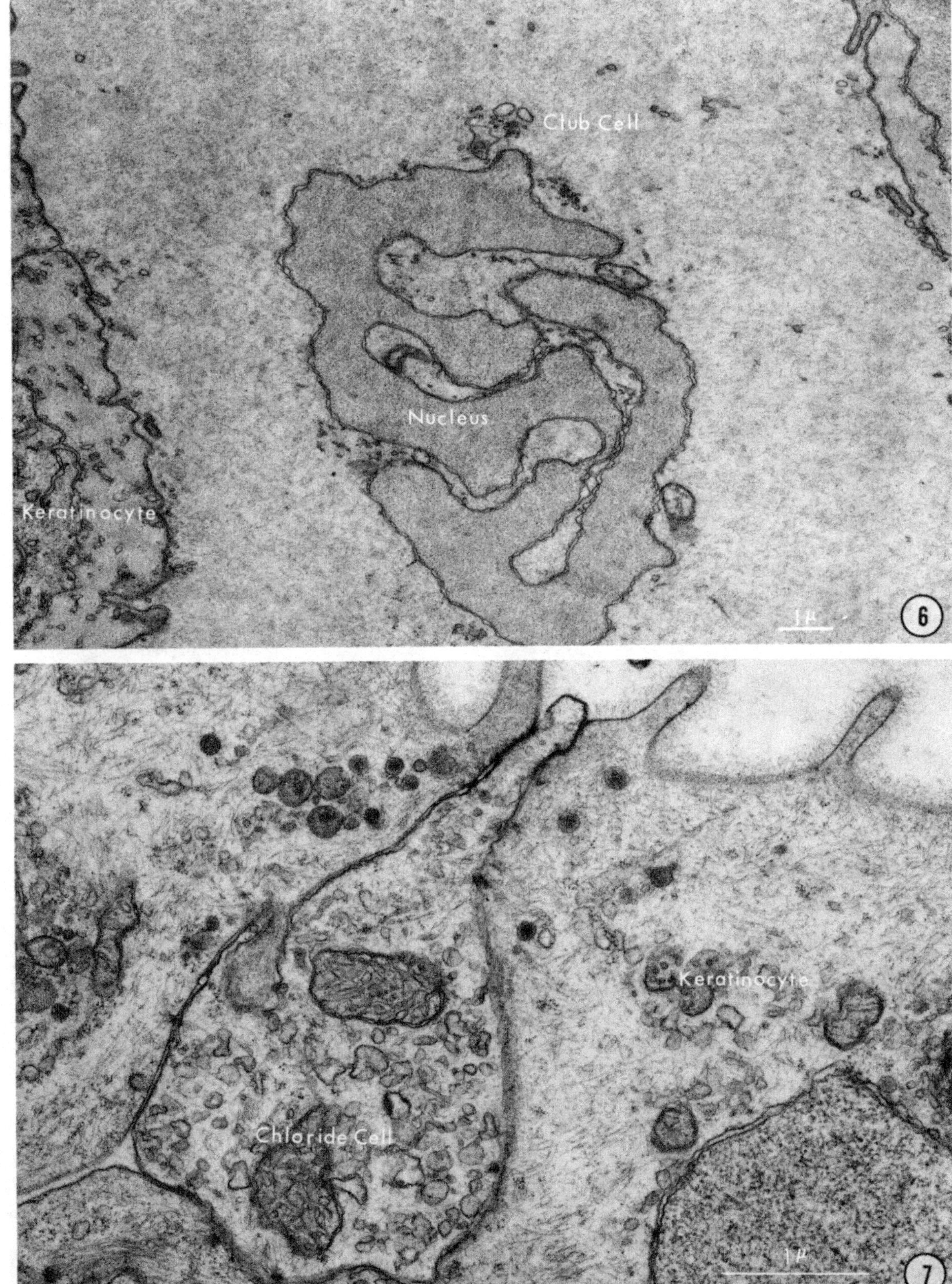

Club Cell
Nucleus
Keratinocyte
1 µ
6
Keratinocyte
Chloride Cell
1 µ
7

III. AMPHIBIAN EPIDERMIS

Amphibians were the first vertebrates to escape a completely aqueous environment; consequently their epidermis has undergone several important modifications. Unlike fish epidermis, amphibian epidermis does not contain chloride or club cells. The main cell type of amphibian epidermis is keratinocyte (Fig. 8) interspersed with a sparse population of mucous cells (Leydig cells). Depending on the species, the superficial layers of the epidermis are either partially or completely keratinized. Toads, for example, are better adapted for a terrestrial life and have a thick cornified layer. In any case, the keratinized layer in amphibians is not dehydrated as it is in reptiles, birds, and mammals; instead, a thin layer of mucus produced by the epidermis and the dermal glands evenly covers the superficial cells. This mucous layer binds water and aids in respiration through the skin.

In contrast to the relatively straight dermo-epidermal junction of fish, that of the frog is highly convoluted (Fig. 13). Following the contours of the basal cells is a well-defined basal lamina about 800 Å thick. Many anchor filaments occur along its entire length joining it with the dermis. Below the basal lamina, collagen fibers are scattered throughout the connective tissue without any well-defined orientation.

The epidermis of *Rana pipiens* consists of 5 to 9 layers of cells (Fig. 9). The basal cells have numerous filaments about 80 Å in diameter. Interspersed among the filaments are mitochondria, small vesicles, and clusters of ribosomes. As these cells migrate upward, they flatten and synthesize various products. Two types of membrane-bounded mucous granules are formed in

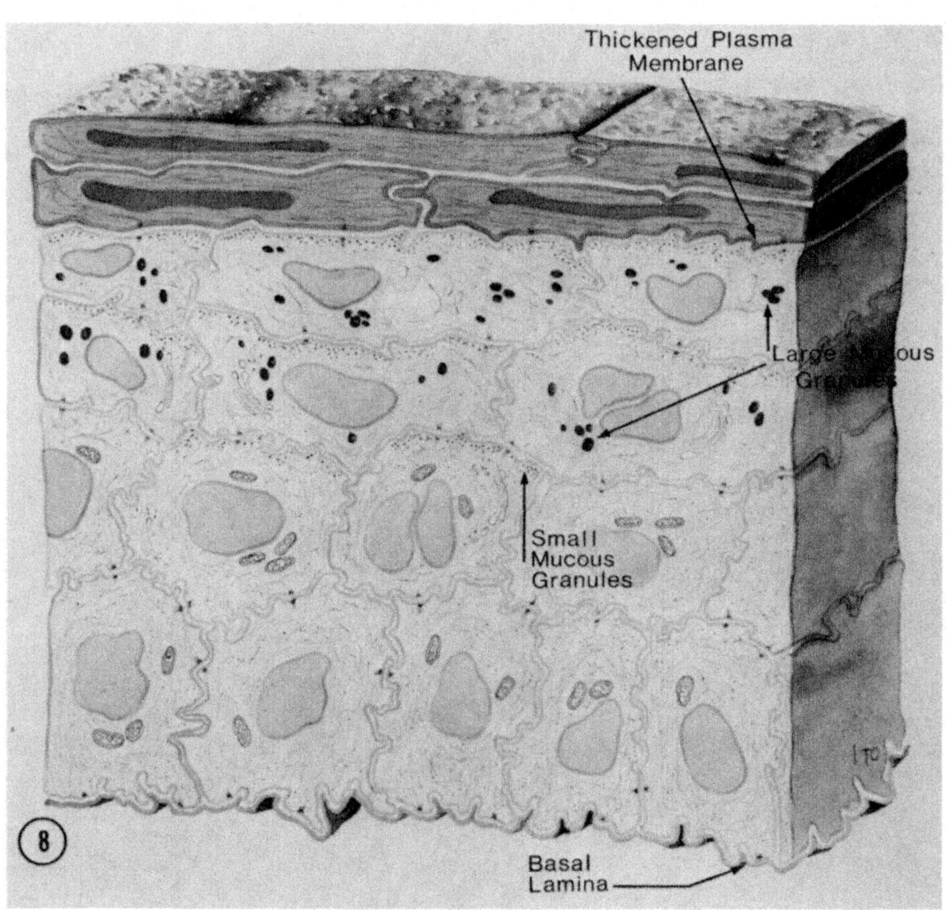

Fig. 8. Schematic diagram of amphibian epidermis.

Fig. 9. Differentiating and superficial cells of frog epidermis (*Rana pipiens*). Veronal acetate buffered osmium fixation. ×11,200.

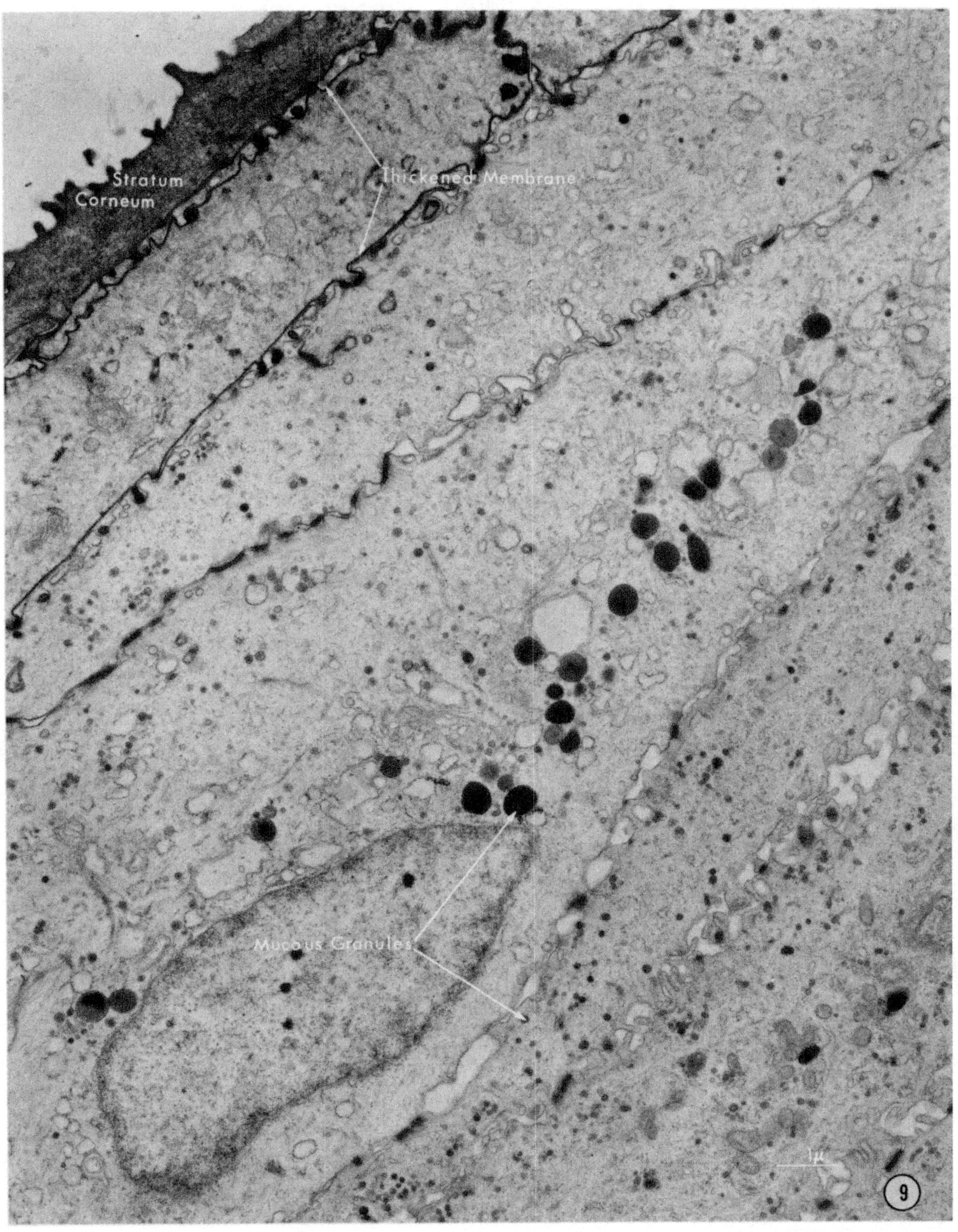
Stratum
Corneum
Thickened Membrane
Mucous Granules
1μ
9

the differentiating cells. The smaller granules (0.04 to 0.08 μ) are formed first and migrate to the apical cell periphery (Fig. 11); many of them are eventually extruded into the intercellular spaces. The larger granules (0.1 to 0.5 μ) are generally located near the nucleus, often in association with the Golgi cisternae (Fig. 10). Numerous vesicles of various sizes in and around the Golgi area contain mucus-like substances that are probably synthesized in this organelle.

During the final stages of differentiation, the nucleus and cytoplasmic organelles are broken down and resorbed and the most superficial cells are filled with oriented 80 Å thick filaments. Since these cells are positive for PAS reaction and since there are no intact mucous granules in the superficial cells, it appears that the mucus is finely dispersed and forms the matrix around the filaments (Fig. 12).

Among the vertebrates, the thickened plasma membranes are first encountered in the superficial cells of the frog epidermis. These cell membranes are about 160 Å in thickness and formed by the deposition of material as a rim on the inner leaflet of the plasma membrane. A thin layer of mucus covers the superficial cells.

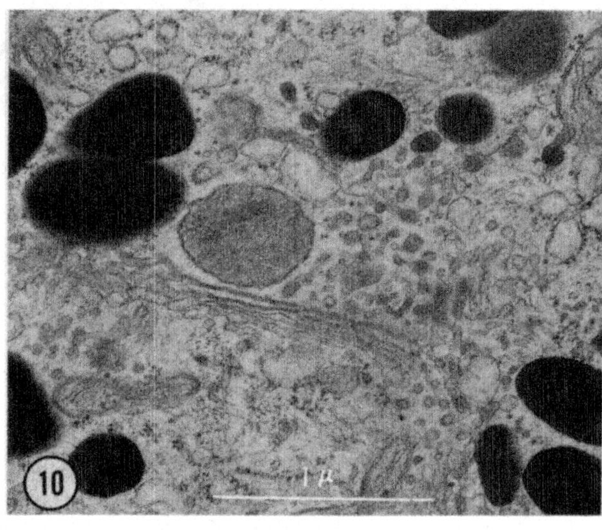

Fig. 10. Production of mucous granules by the Golgi complex. Veronal acetate buffered osmium fixation. ×30,000.

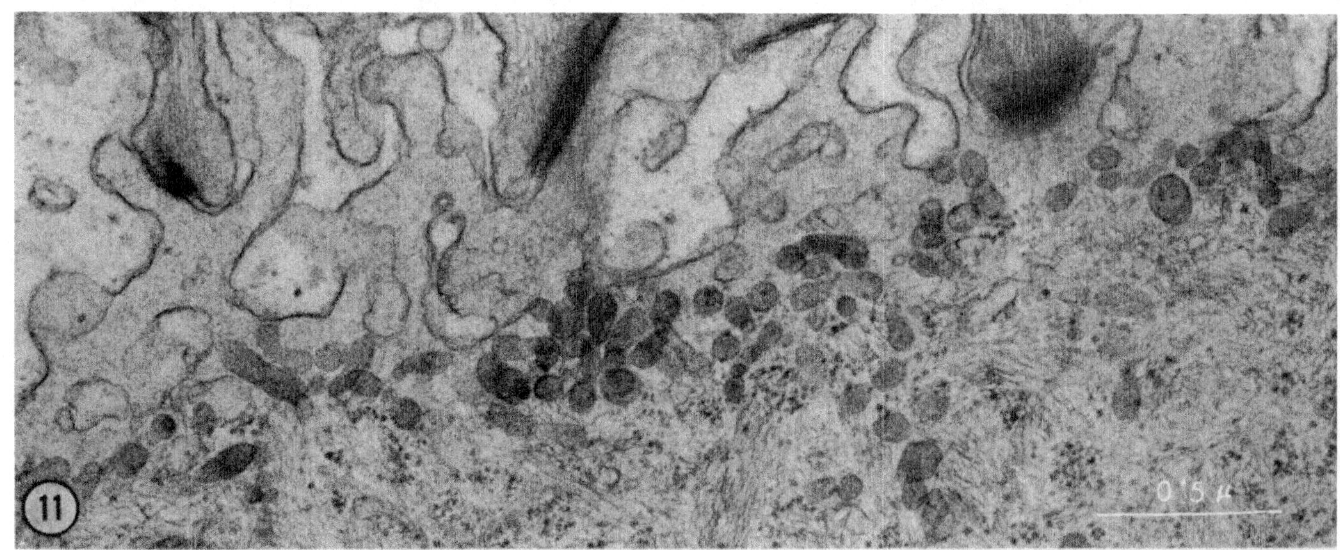

Fig. 11. Alignment of small mucous granules along apical cell periphery. Veronal acetate buffered osmium fixation. ×50,000.

Fig. 12. Superficial keratinized cells filled with filaments showing the thickened plasma membrane. Veronal acetate buffered osmium fixation. ×22,000.

Fig. 13. Convoluted dermo-epidermal junction of frog skin (*Rana pipiens*). Veronal acetate buffered osmium fixation. ×18,000.

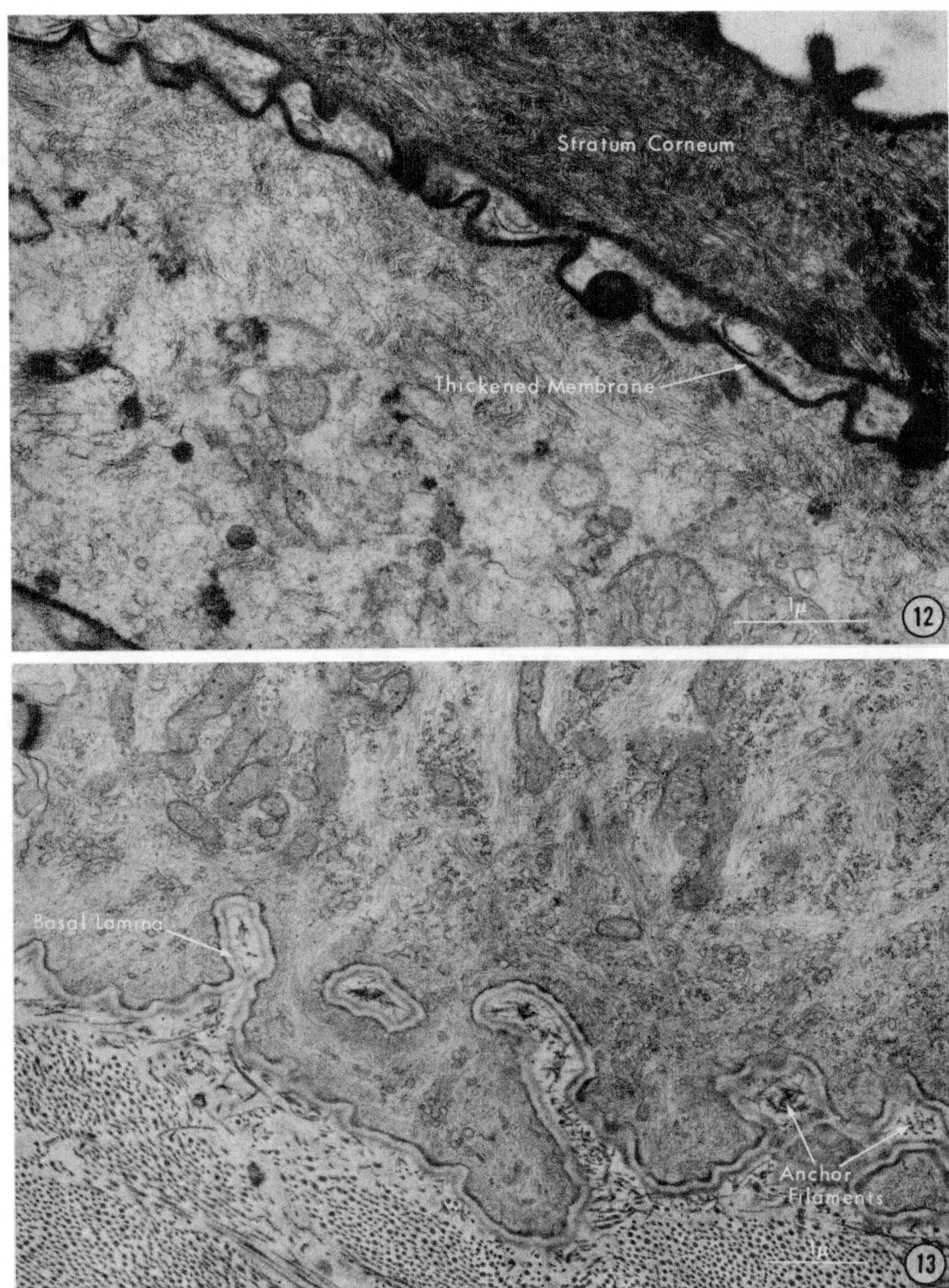
Stratum Corneum
Thickened Membrane
1 μ
12
Basal Lamina
Anchor
Filaments
1 μ
13

IV. REPTILIAN EPIDERMIS

The first completely terrestrial vertebrates to evolve had various adaptations to insure their survival in a dry environment. Their characteristic horny scales, lightweight and flexible and often overlapping, prevent desiccation and provide a protective armor. The scales are composed of distinct layers; the two most prominent are called α and β layers because of their x-ray diffraction pattern. Electron microscope studies show that the cells of the α layer are mainly filled with 80 Å filaments, those of the β layer with 30 Å filaments. The distribution of these two layers varies within the different reptilian orders: α-keratin is indigenous to moveable, stretchable areas, β-keratin to inflexible places where its main function is protection.

Lizards and snakes have the most complex arrangement of epidermal layers, which are periodically renewed by molting. During the resting stage, four layers of cornified cells lie above a stratum of living cells as exemplified by Figs. 14 and 15, which show the resting epidermis from the back of *Anolis carolinensis*. The superficial layer, called the Oberhäutchen, is a thin covering with many tapering spinules. The thick cornified β layer is without apparent cell boundaries but is conspicuously patterned and packed with 30 Å filaments (upper inset of Fig. 15). Relatively inflexible, it is several microns thick in the center of the scale but tenuously thin at the hinge regions. Beneath the β layer is the mesos or intermediate layer, characterized by flattened cells that lack desmosomal plaques. During preparation and sectioning, the epidermis in this region often separates. Underneath the mesos lies the α layer of cells filled with 80 Å light filaments surrounded by an electron opaque matrix (lower inset of Fig. 15). The cell boundaries of the α layer are clearly visible in contrast to those of the β layer. During the resting stage, the living stratum beneath the α layer consists of one to two cell layers. The mature Oberhäutchen, β, mesos, and α layers, as well as these living cells, constitute an epidermal generation.

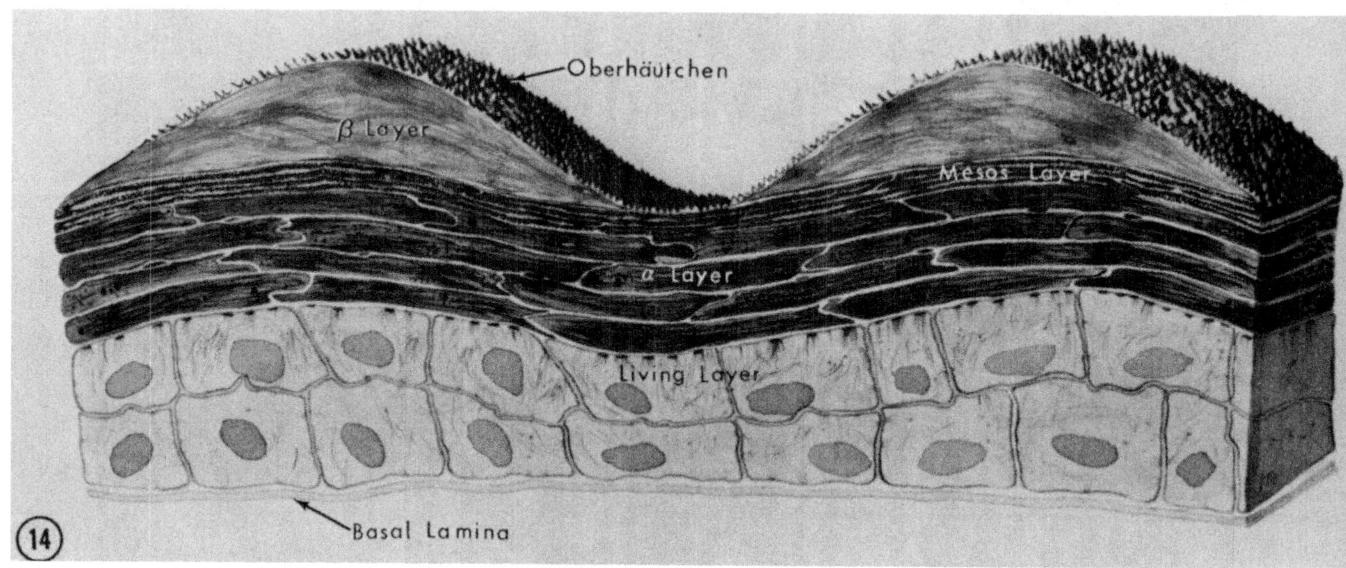

Fig. 14. Diagram of resting lizard epidermis. [From Alexander, N.J. (1970). *Z. Zellforsch.* **101**, 156. Courtesy of the publisher.]

Fig. 15. Resting stage epidermis of the American chameleon (*Anolis carolinensis*) showing the relationship of the Oberhäutchen, β, mesos, and α layers. Collidine buffered osmium fixation. ×13,000. [From Alexander, N.J., and Parakkal, P. F. (1969). *Z. Zellforsch.* **101**, 75. Courtesy of the publisher.]

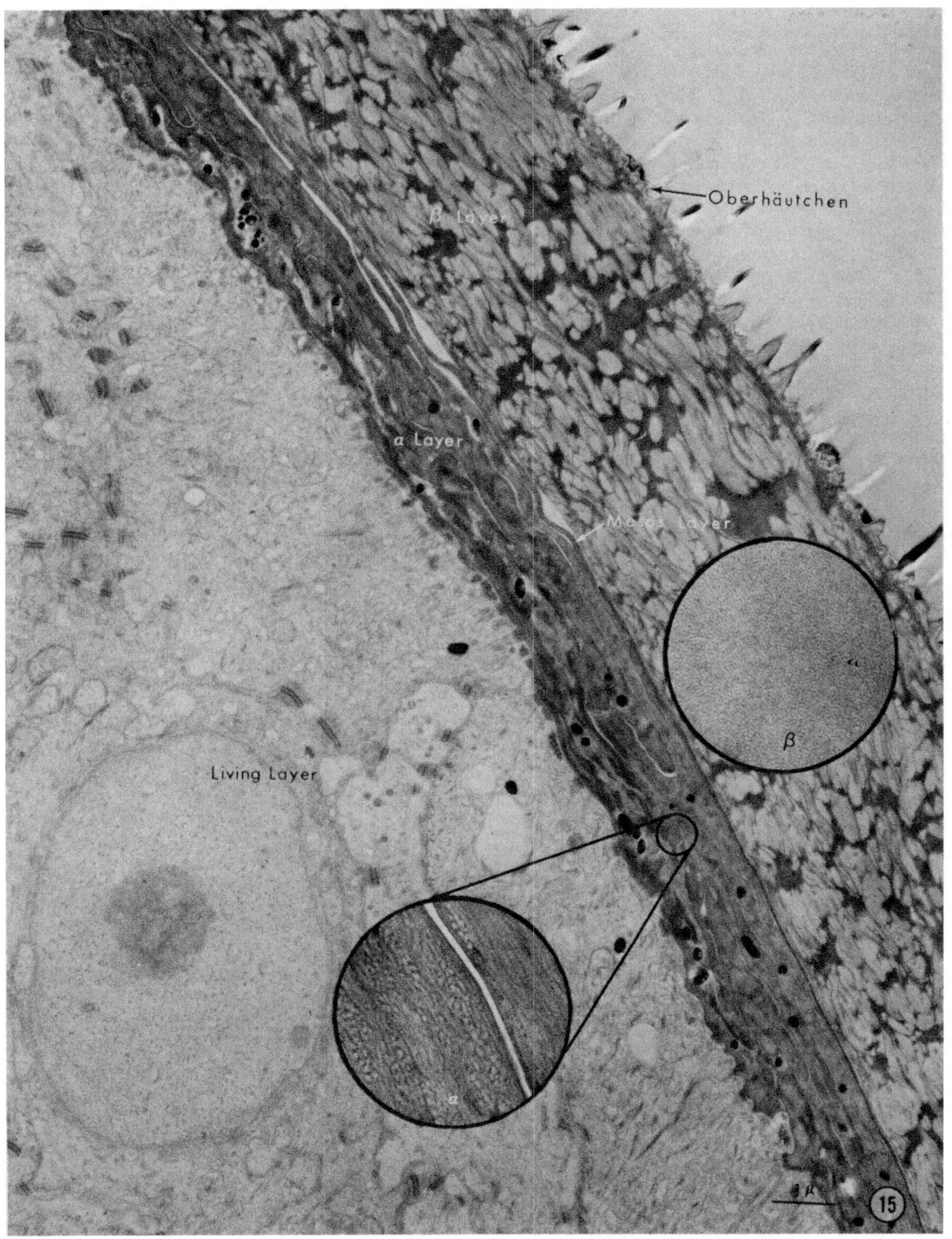
Oberhäutchen
β Layer
α Layer
Mesos Layer
Living Layer
β
α
1 μ
15

A. Squamates

Before the molt in squamates, a new set of epidermal layers (Figs. 16 and 17) is formed underneath the old generation (Fig. 15). The harbinger of an approaching molt is the differentiation of a single layer of cells that eventually becomes the outer layer of Oberhäutchen. These forming cells produce spinules that interdigitate with the basal cells of the overlying epidermal generation. Within the spinules, 80 Å filaments accumulate and interconnect with the bases of adjacent filament groups. The new Oberhäutchen is firmly attached to the overlying layer by means of desmosomes at the base of the spinules. When the Oberhäutchen becomes quite distinct, the adjacent cells begin to differentiate into the β layer. Small membrane-bounded packets containing an amorphous material are formed within the presumptive β cells. As differentiation continues, these packets coalesce until the whole cell is filled except for scattered accumulations of glycogen. The mesos layer, which exhibits the characteristics of both the α and β layer, is not seen in this micrograph. The presumptive α layer, like the mesos, can be recognized at this time only by its position below the forming β layer. α Layer formation begins with accumulations of glycogen (inset) in the cytoplasm and is followed by aggregations of wispy filaments about 80 Å in thickness. Later the filaments form bundles and fill the cells to the exclusion of other organelles. When fully cornified, the cells of the α layer have a typical "keratin pattern," an arrangement of electron-lucent filaments surrounded by an opaque matrix (see lower inset of Fig. 15).

Above the forming epidermal generation is a layer of cells called the clear layer, which plays an important role in the shedding of the old epidermal generation. As the time of molting approaches, striking changes occur in this layer. Keratohyalin-like material accumulates in large clumps and eventually obscures the cellular organelles. Before molting, desmosomal attachments appear and the skin is shed at the junction of the clear cells and the forming Oberhäutchen.

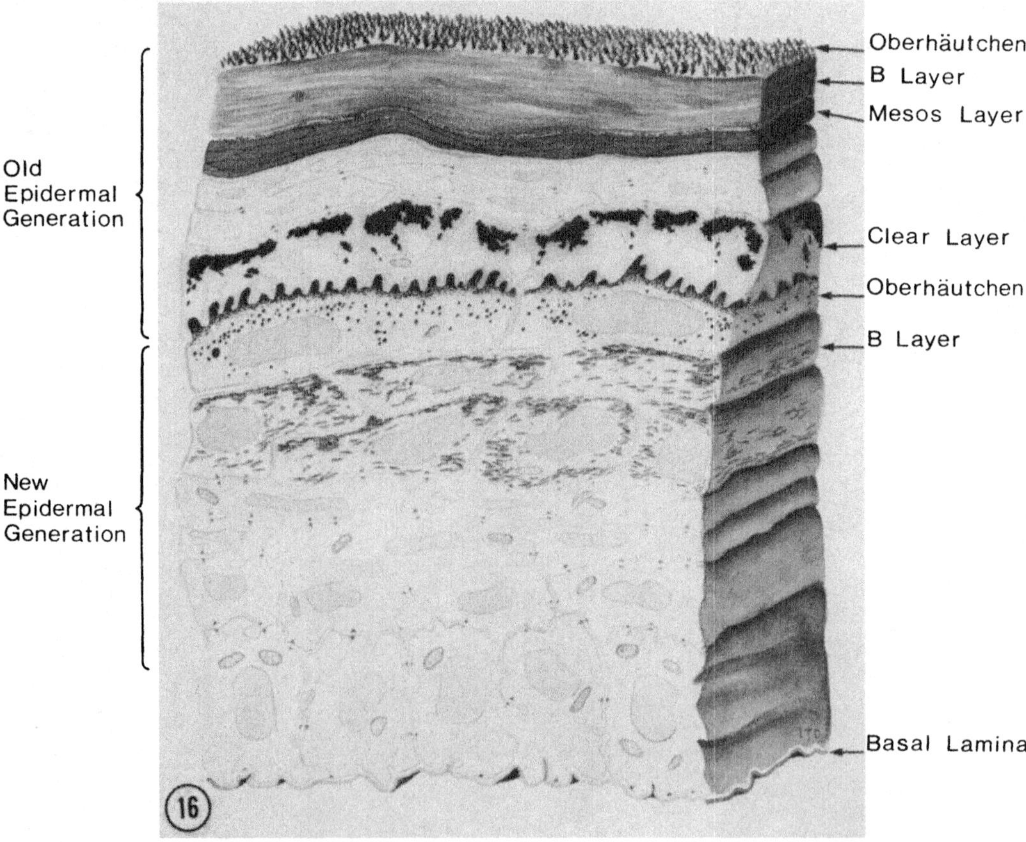

Fig. 16. Schematic diagram of molting lizard skin, showing the formation of a new epidermal generation under the old.

Fig. 17. Molting belly epidermis of the American chameleon (*Anolis carolinensis*) showing mainly the new epidermal generation. Collidine buffered osmium fixation. ×9000. [From Alexander, N.J., and Parakkal, P. F. (1969). *Z. Zellforsch.* **101**, 76. Courtesy of the publisher.]

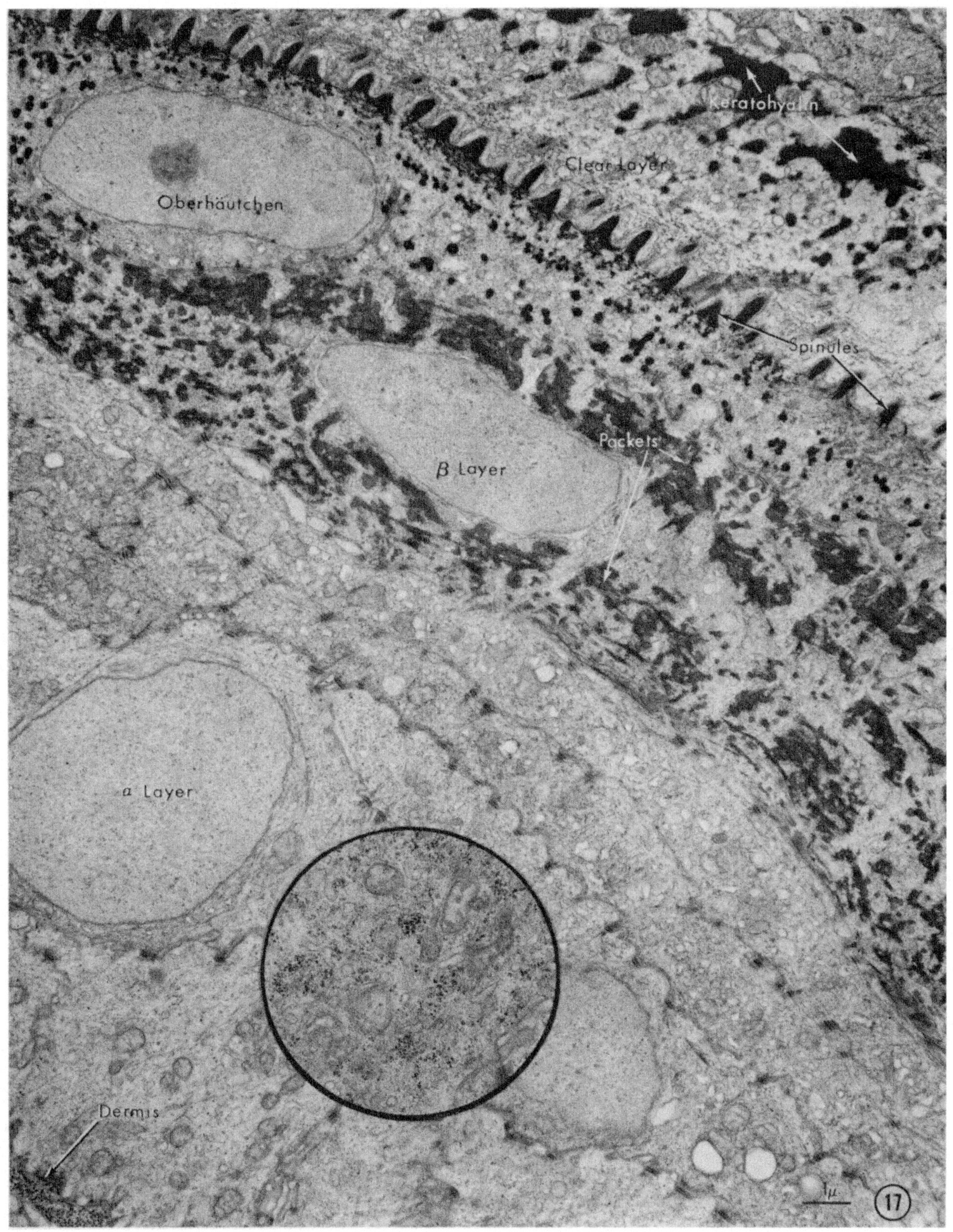
Oberhäutchen
Clear Layer
keratohyalin
Spinules
β Layer
Packets
a Layer
Dermis
1μ
17

B. Chelonians

The distribution of the α and β layers varies in the different orders of reptiles. The shell of turtles and tortoises has only a β layer which overlays the bony carapace (dorsal) and plastron (ventral) (Fig. 18). Because there is no series of different layers, molting *in toto* does not occur; instead small patches of cornified cells are sloughed off from time to time. The β layer, which grows around the periphery of the present layer, produces the rings seen on the epidermis of the shell of some chelonians, particularly tortoises.

The integument of the carapace of the red-eared terrapin, *Pseudemys scripta,* has a typical β layer (Fig. 19). The boundaries between the cornified cells coalesce just as in the β layer of squamates and the entire layer appears reticulated, the darker lines being at least partially formed from tonofilaments and associated desmosomes. The junction between the cornified cells and the living stratum is convoluted and held in place by means of many desmosomes. The living cells directly beneath the cornified layer often possess packets of electron-lucent material which accumulates and becomes enmeshed in 80 Å filaments. These packets become the less dense areas of the β layer and yield 30 Å fibrillar structures whereas the more dense areas are formed from the 80 Å filaments.

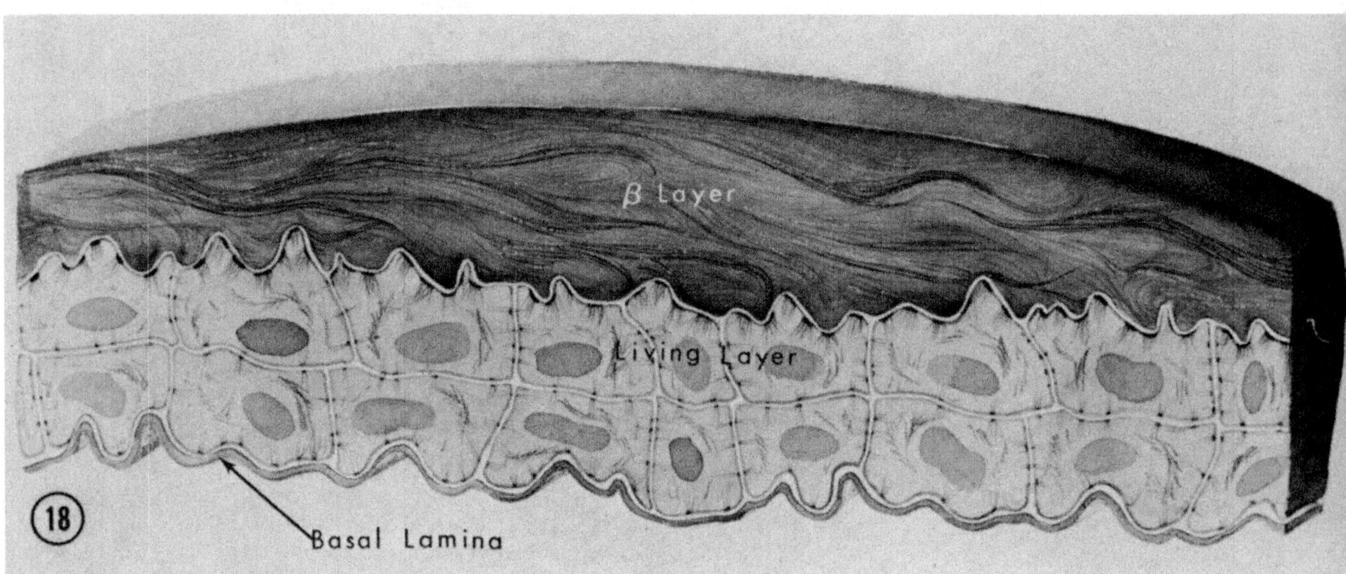

Fig. 18. Diagram of a turtle shell. [From Alexander, N.J. (1970). *Z. Zellforsch.* **110,** 156. Courtesy of the publisher.]

Fig. 19. Junction of the β layer of the red-eared terrapin (*Pseudemys scripta*) with the underlying living cells. Collidine buffered glutaraldehyde, osmium postfixation. ×56,000. [From Alexander, N.J. (1970). *Z. Zellforsch.* **110,** 159. Courtesy of the publisher.]

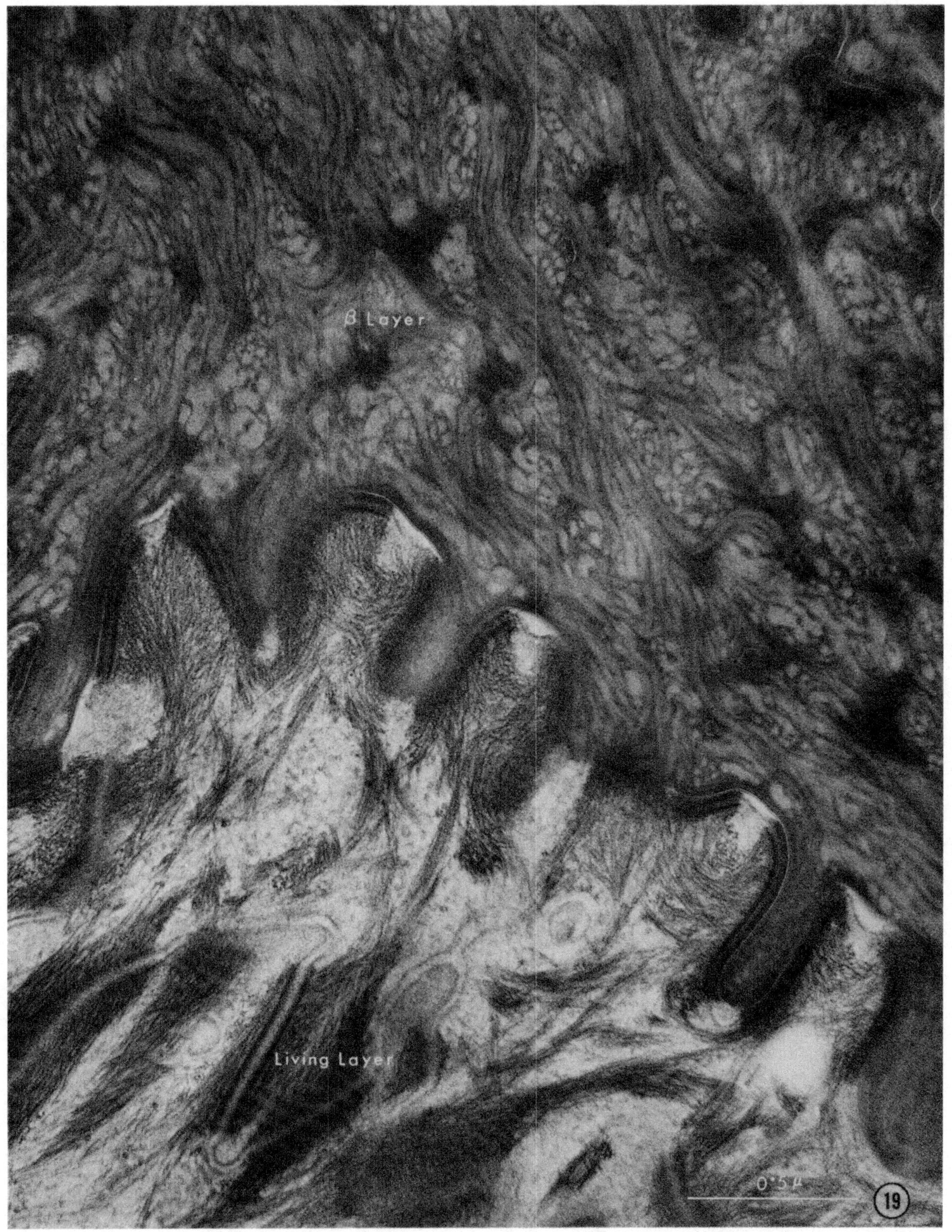
β Layer
Living Layer
0·5 μ
19

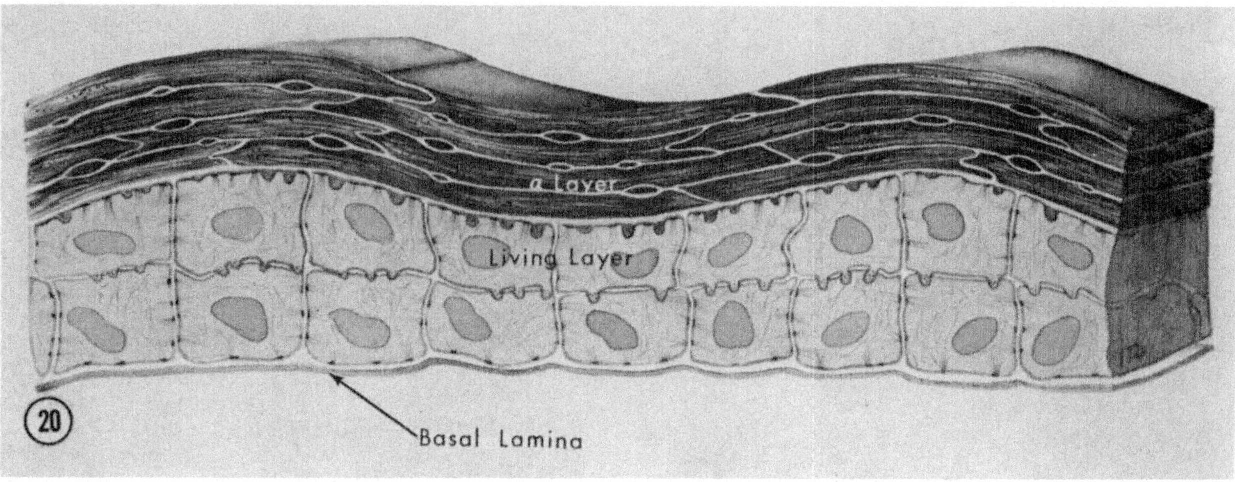

Fig. 20. Diagram of turtle leg epidermis. [From Alexander, N.J. (1970). *Z. Zellforsch.* **110,** 156. Courtesy of the publisher.]

Fig. 21. Keratinized α layer and the living layer beneath from the leg epidermis of the map turtle (*Graptemys geographica*). Collidine buffered glutaraldehyde, osmium postfixation. $\times 43,000$.

The flexible skin of the leg (Figs. 20 and 21), head, and neck of turtles and tortoises contains only an α layer. This arrangement of a cornified α layer over a region of living cells is similar to that in mammalian skin. The cornified cells of the α layer retain distinct boundaries. Interspersed among the desmosomal remnants are accumulations of PAS-positive material that arises in vesicles within the cells of the living stratum. The vesicles discharge at the superficial cell surface and concomitantly the plasma membrane of the cornifying cells thickens. A second type of granule accumulates in association with 80 Å filaments in the center of the cornifying cells. These multigranular bodies are 0.5–2 μ in diameter and are membrane-bounded. Often these bodies are composed of several smaller packets consisting of alternating light and dark lamellae. These granules are similar to those present in bird epidermis and likewise they are not expelled but become incorporated within the cornifying cell.

C. Crocodilians

The epidermis of alligators and caimans has only a single cornified layer of cells which varies in composition in different parts of the scale (Fig. 23). Figure 24 of the back epidermis of the spectacled caiman, *Caiman sclerops*, is from an area near the periphery of a scale. The cornified layer has many of the characteristics of the lizard α layer, such as definite cellular

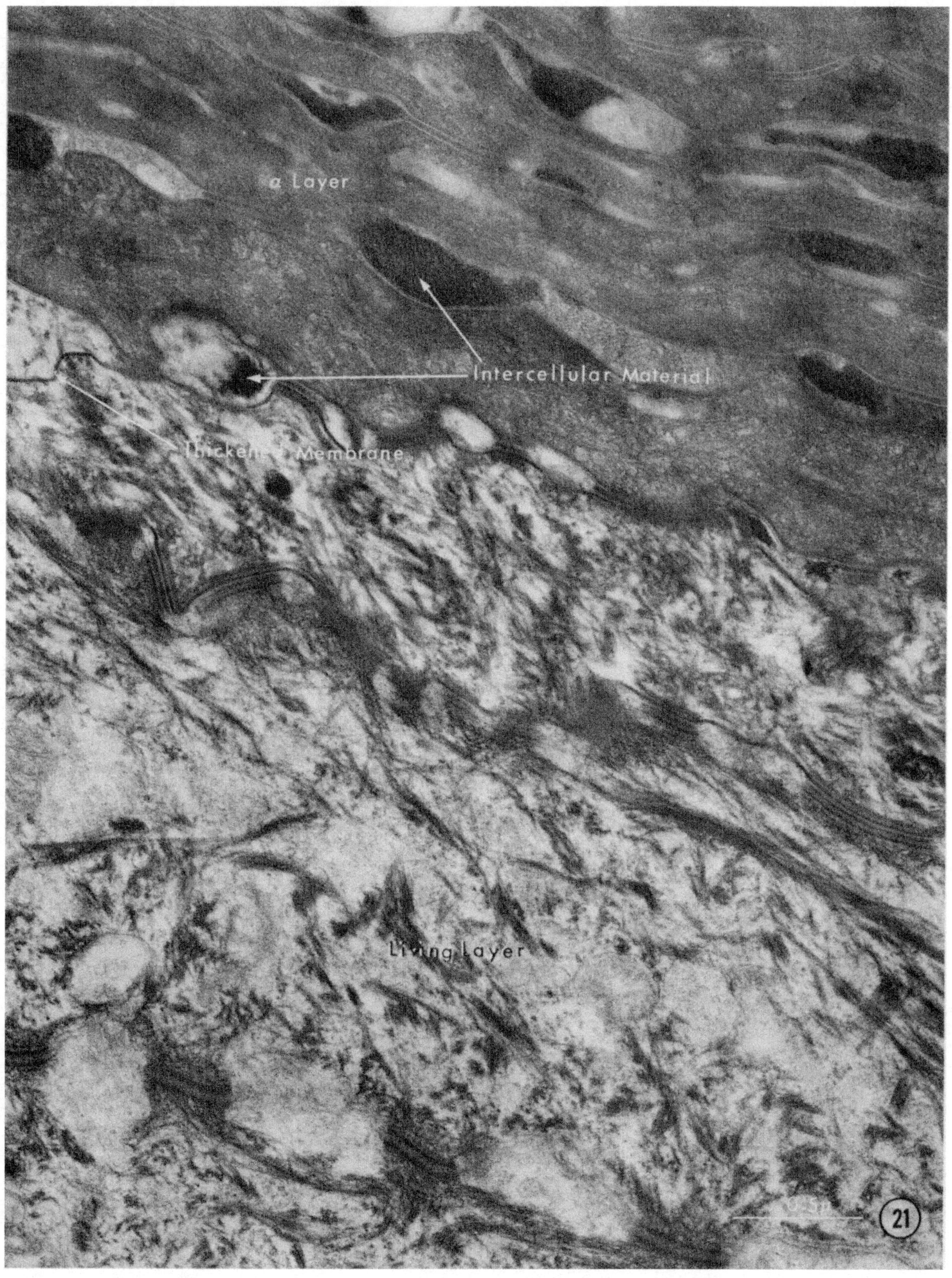

a Layer
Intercellular Material
Thickened Membrane
Living Layer
21

outlines, desmosomal plaques, and packets of extracellular mucopolysaccharide, but is probably a mixture of α and β into a combination layer. The living cells beneath the cornified layer are unique among reptiles because only in crocodilians do these cells contain clumps of keratohyalin that accumulate around the cell periphery in association with bundles of 80 Å filaments. The center of the cell fills with multigranular bodies (Fig. 22), similar to those in the epidermis of the turtle leg, and with remnants of cell organelles. At a later stage than that illustrated here, the plasma membranes of the cornifying cells of the hinge region thicken like those in the α layer of turtle leg and mammalian epidermis.

Cells in the center of the scale resemble those of the β layer of lizard skin and turtle shell. They are patterned and without residues of organelles, although they retain the remnants of desmosomes and distinct plasma membranes that do not thicken during cornification.

The different rhythms and patterns of keratin formation that have evolved within the orders of reptiles are all variations of a common process. In squamates, a sequence of layers is formed periodically; the lowest or clear layer of the old epidermal generation cornifies and the skin is molted. At the other end of the spectrum, the scales of chelonians and crocodilians, composed of either α or β layers, are keratinized and shed in a slow continuous process like that in bird and mammalian epidermis. Turtles and crocodilians share two additional characteristics with birds and mammals: (a) the plasma membrane of the α layer thickens; (b) the α layer contains deposits of extracellular mucopolysaccharides. Only in crocodilians does one find a granular layer with accumulations of keratohyalin. Thus, despite their numerous unique characteristics, the scales of reptiles possess some common features with the epidermis of other classes of vertebrates.

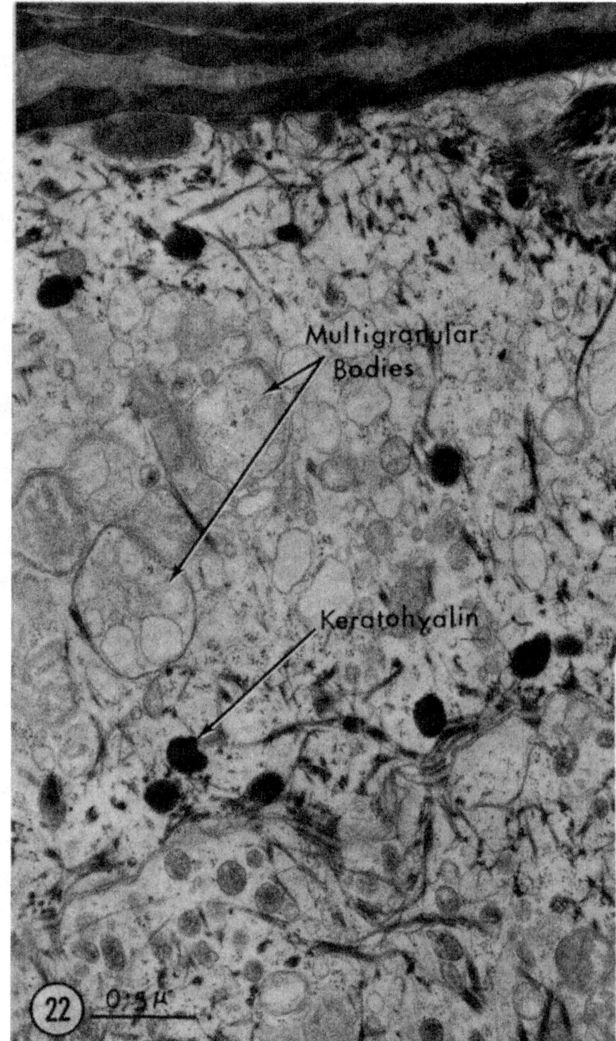

Fig. 22. A portion of a differentiating cell of caiman epidermis showing keratohyalin and multigranular bodies. Collidine buffered osmium fixation. ×36,000.

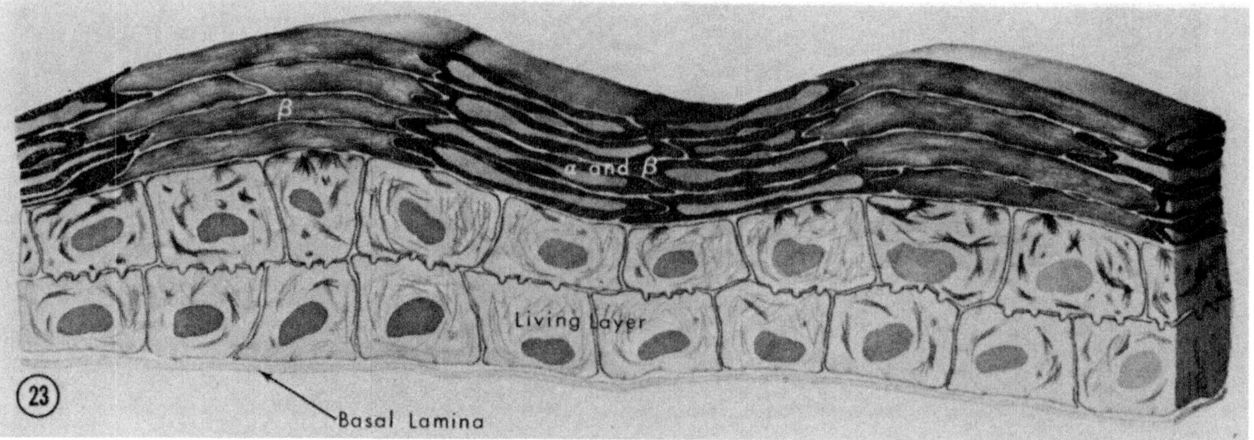

Fig. 23. Diagram of crocodilian epidermis. [From Alexander, N.J. (1970). *Z. Zellforsch.* **110,** 156. Courtesy of the publisher.]

Fig. 24. Back epidermis of the spectacled caiman (*Caiman sclerops*). Collidine buffered osmium fixation. ×24,000.

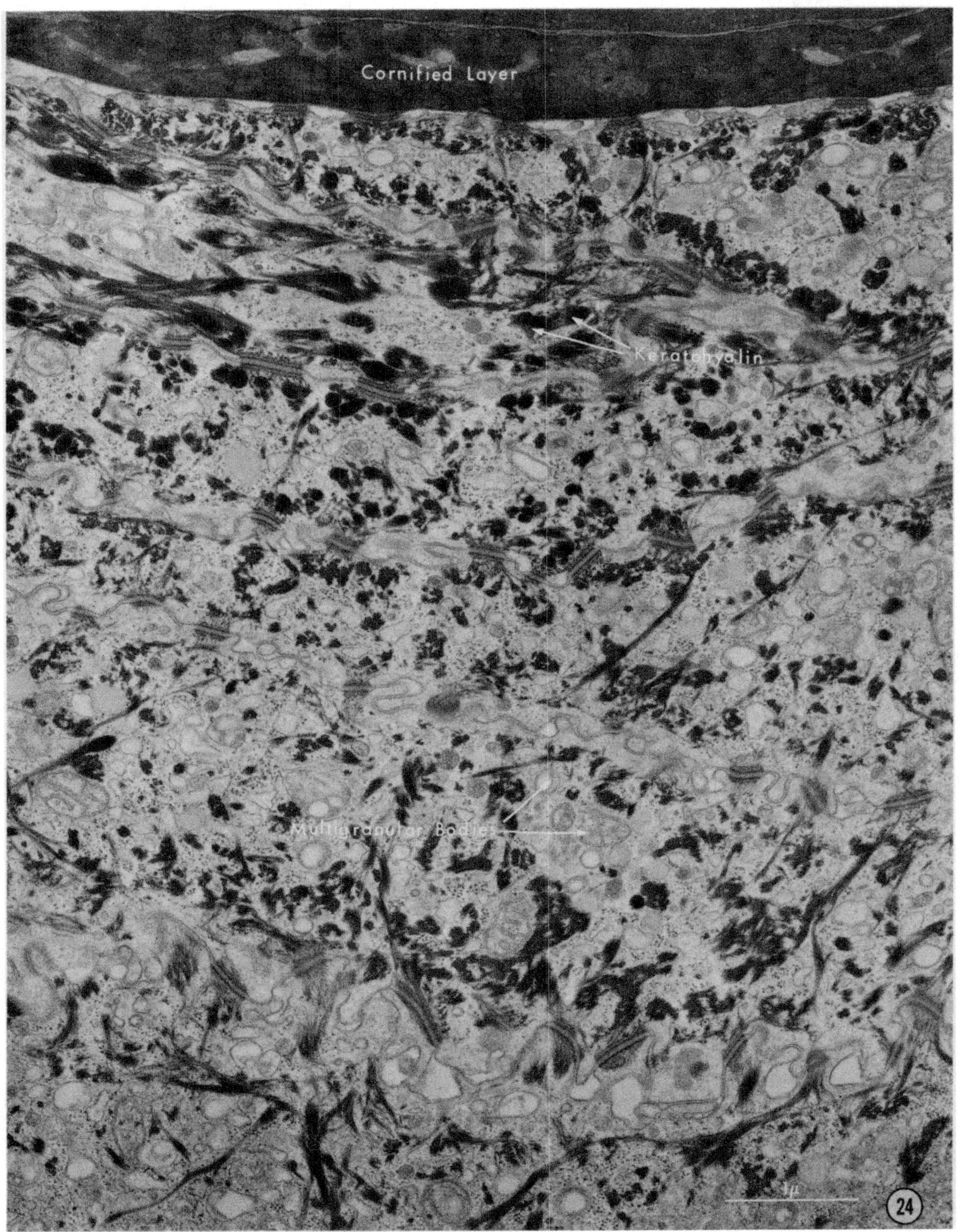
Cornified Layer
Keratohyalin
Multigranular Bodies
1μ
24

V. AVIAN EPIDERMIS

Many of the keratinized structures of birds, such as scales, claws, and the β-keratin of feathers, reflect their reptilian ancestry. The avian epidermis, even though it shares many of the common features of keratinizing epithelia, nonetheless differs from both reptilian and mammalian epidermis.

The avian epidermis is composed of a stratified squamous epithelium that is continuously renewed. The basal cells divide, migrate upward, and differentiate to form the horny cells. During differentiation many cytoplasmic filaments 80 Å in diameter are formed similar to those in the keratinocytes of fish, amphibia, and mammals, and in the α layer or reptilian epidermis. Large quantities of lipids not usually found in vertebrate epidermis are formed in bird epidermis (Fig. 25). Lipogenesis is accompanied by a hypertrophy of the Golgi and smooth endoplasmic reticulum, reminiscent of lipid synthesis in the sebaceous glands of mammals. Since sebaceous glands are not a normal component of the skin of birds, the epidermis may have taken over some of their functions.

Another characteristic differentiation product both in reptiles and in birds is the multigranular bodies. These ovoid bodies are membrane bounded and when fully formed are 0.5 μ in diameter. They contain from two to six lamellated granules that are not membrane bounded (Fig. 26, inset). These granules are incorporated into the substance of the fully cornified cells; their function is obscure.

During their differentiation, the cells also accumulate numerous small keratohyalin granules in close association with the 80 Å filaments near the cell periphery. Since these keratohyalin granules are small, their presence in the avian epidermis has invited some controversy. Electron microscope studies have demonstrated that keratohyalin granules are normal components of the differentiating avian epidermal cells (Fig. 26). Except for their small size, they are identical morphologically with those in mammalian epidermis.

During the final stages of differentiation, the nuclei and cytoplasmic organelles are decomposed and resorbed. Simultaneously the plasma membrane undergoes the thickening common to all keratinizing cells. The completely cornified cells are extremely flat and filled with filaments surrounded by an amorphous matrix.

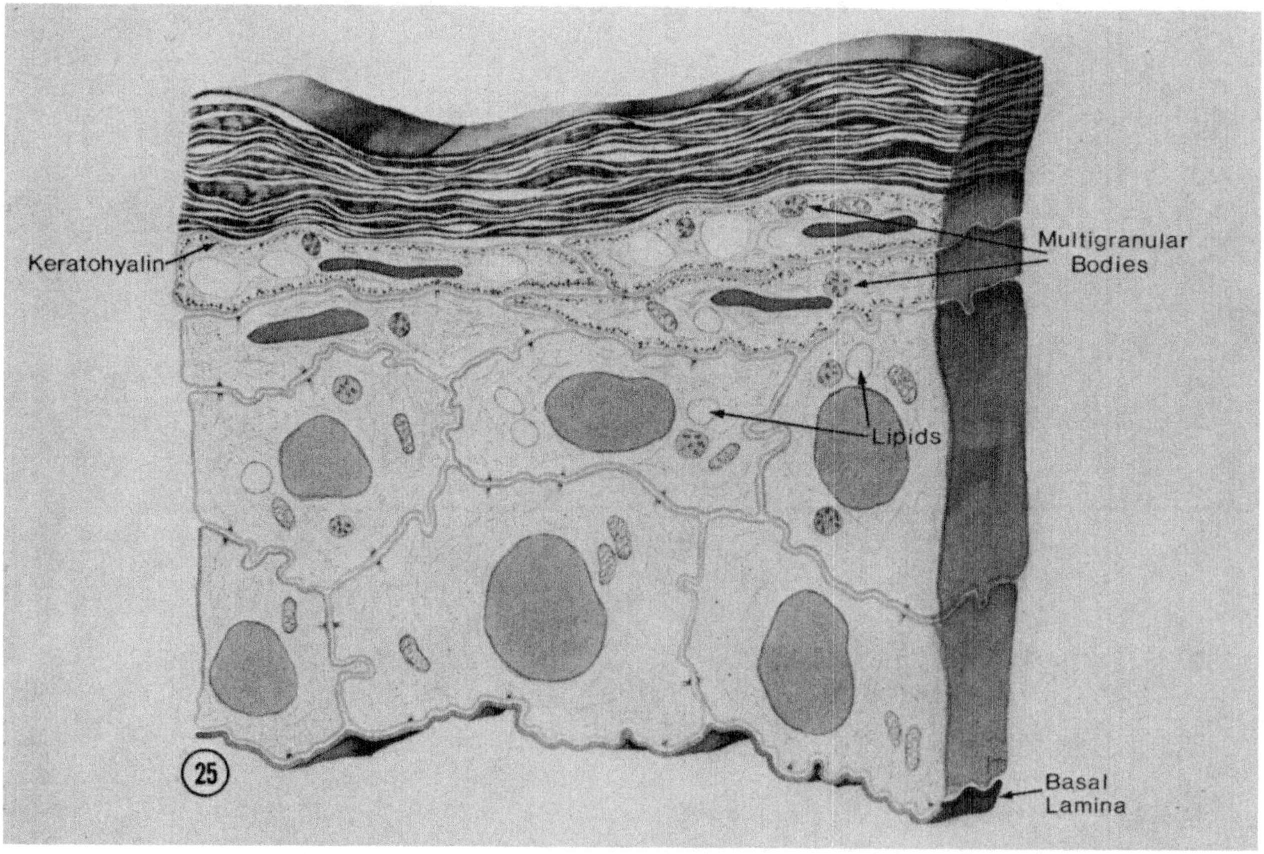

Fig. 25.   Diagram of adult bird skin.

Fig. 26.   Epidermis of adult chicken showing lipid droplets, multigranular bodies (inset), and keratohyalin granules. Veronal buffered osmium fixation. ×25,500.

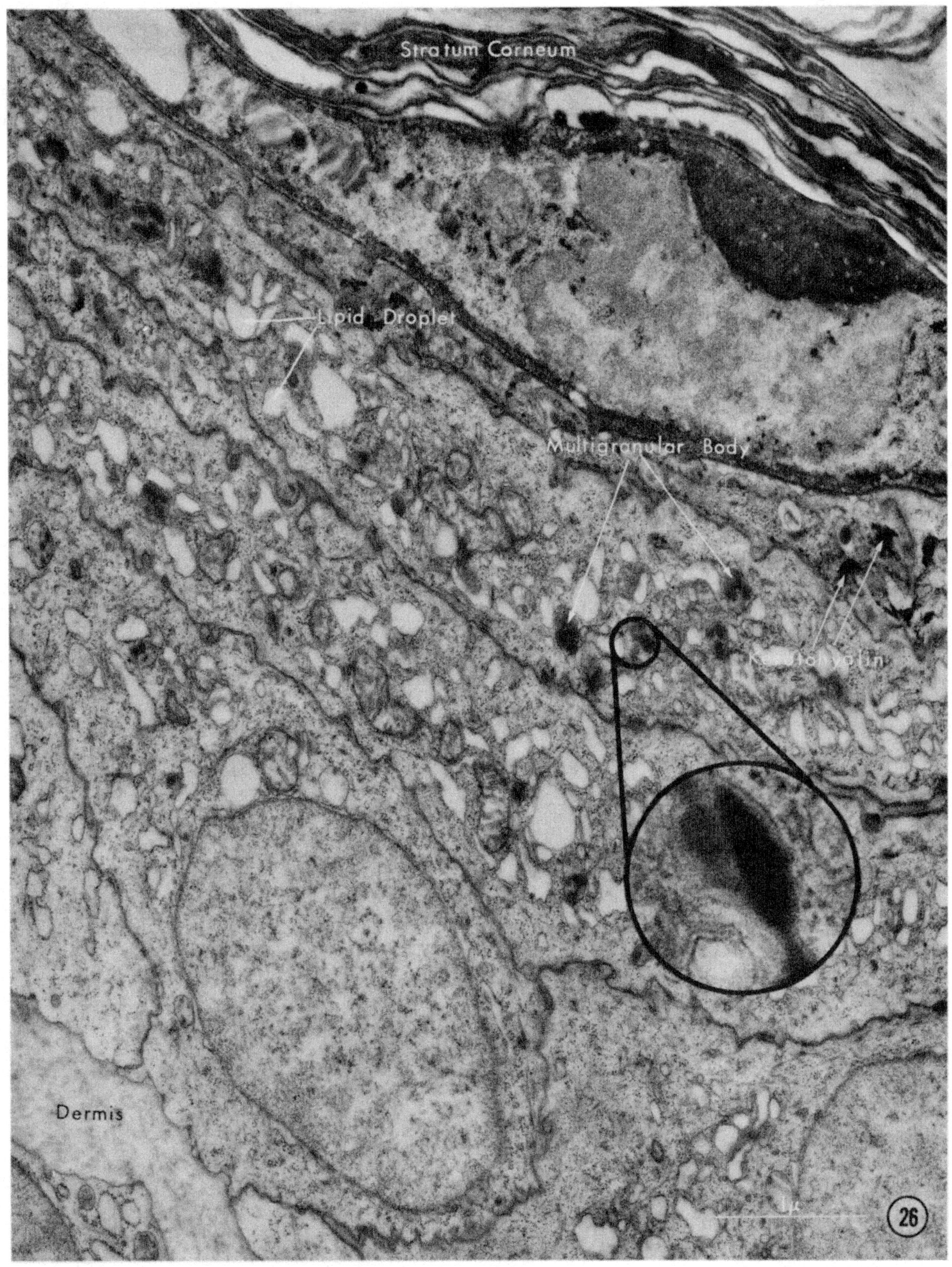

Stratum Corneum
Lipid Droplet
Multigranular Body
Keratohyalin
Dermis
26

A. Development

Chick skin is a good model for developmental studies, especially because of the availability of embryos of known age. During early development, the covering epithelium is composed of only one layer of flattened cells. As the chick develops, this stratum becomes bilayered, consisting of a columnar basal cell and a squamous superficial cell. Eventually the columnar cell gives rise to the definitive epidermis and the superficial cell forms the periderm. In a 7-day-old chick embryo, the covering epithelium consists of only three layers of cells (Fig. 27). The basal columnar cells are separated from the underlying dermis by a basal lamina. These cells characteristically contain large numbers of ribosomes either singly or in polysomes, some rough-surfaced endoplasmic reticulum, and well-formed Golgi areas. Only a few desmosomes are seen among the cells which often have large intercellular spaces. The superficial layer is the presumptive periderm and can be recognized only by its position.

By the time the embryo is 16 days old, the periderm is bilayered and has characteristic differentiation product, the peridermal granules. They are irregular in shape and size and are distributed throughout the cell. The distribution in the two layers of the periderm varies. In the superficial layer of peridermal cells, the granules are fewer and smaller than in the layer underneath (Fig. 28).

By day 16, the epidermis below the bilayered periderm has acquired the structural characteristics of adult epidermis. It consists of three to four layers, but a horny layer has not yet been formed. However, both lipids and multigranular bodies are present in the cells of the upper layers. Numerous keratohyalin granules are also seen in the cells of the uppermost layer. The intercellular junctions, previously scarce, become numerous, occurring at close intervals along the cell membranes. The wide intercellular spaces common in younger embryos are absent.

The periderm is a distinct tissue and follows a characteristic course of development. Probably it protects the epidermis during embryonic development, becoming superfluous when the stratum corneum assumes its protective function. The disintegration and eventual loss of the periderm is probably correlated with the newly formed stratum corneum interfering with the passage of nutrients from the epidermis. Although the periderm is only two layers thick over most of the body, it sometimes reaches six to eight layers over the developing beak.

The most characteristic products of the periderm cells are the granules and the large accumulations of glycogen (Fig. 29). The periderm granules have been mistaken for keratohyalin and trichohyalin because of their common staining properties; however, electron microscopic studies have demonstrated that they are morphologically different from both keratohyalin and trichohyalin. The periderm granules consist of interlacing strands 200–300 Å thick, which enclose spaces between 500–1000 Å wide to produce a mesh-like structure. At high magnification, these strands are made up of particles about 60–80 Å in diameter.

Fig. 27. Skin of a 7-day-old chick embryo showing the presumptive epidermis and periderm. Veronal buffered osmium fixation. ×7000.

Fig. 28. Skin of a 16-day-old chick embryo showing the well-developed bilayered periderm and the underlying granular layer. Veronal buffered osmium fixation. ×10,300. [From Parakkal, P. F., and Matoltsy, A. G. (1968). *J. Ultrastruct. Res.* **23**, 409. Courtesy of the publisher.]

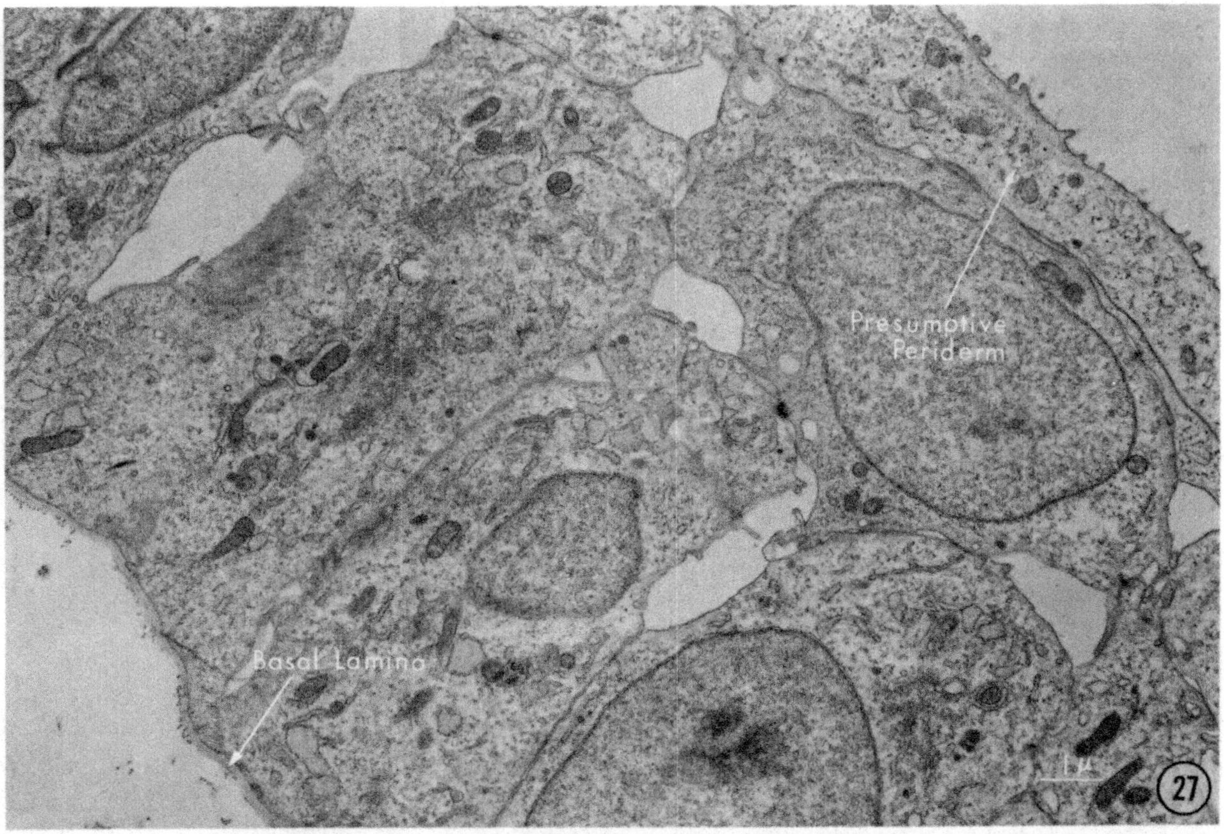

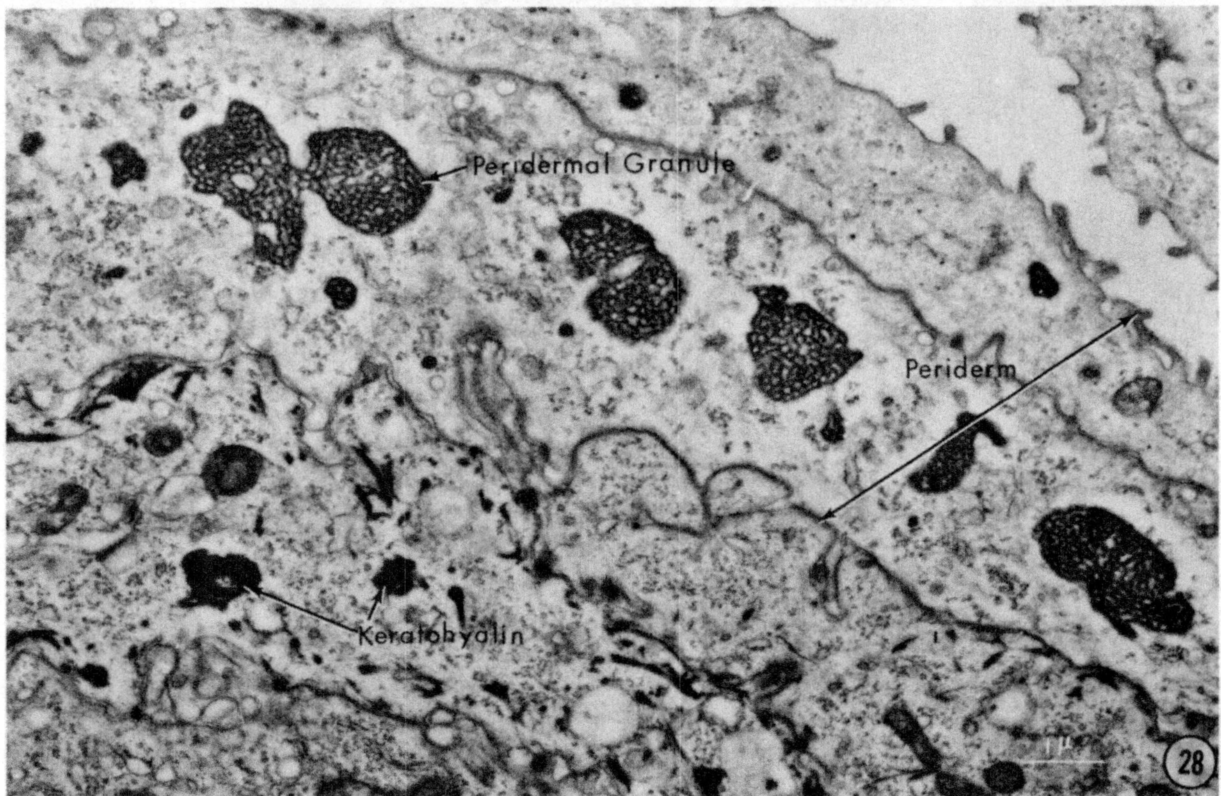
Presumptive
Periderm
Basal Lamina
1 μ
27
Peridermal Granule
Periderm
Keratohyalin
1 μ
28

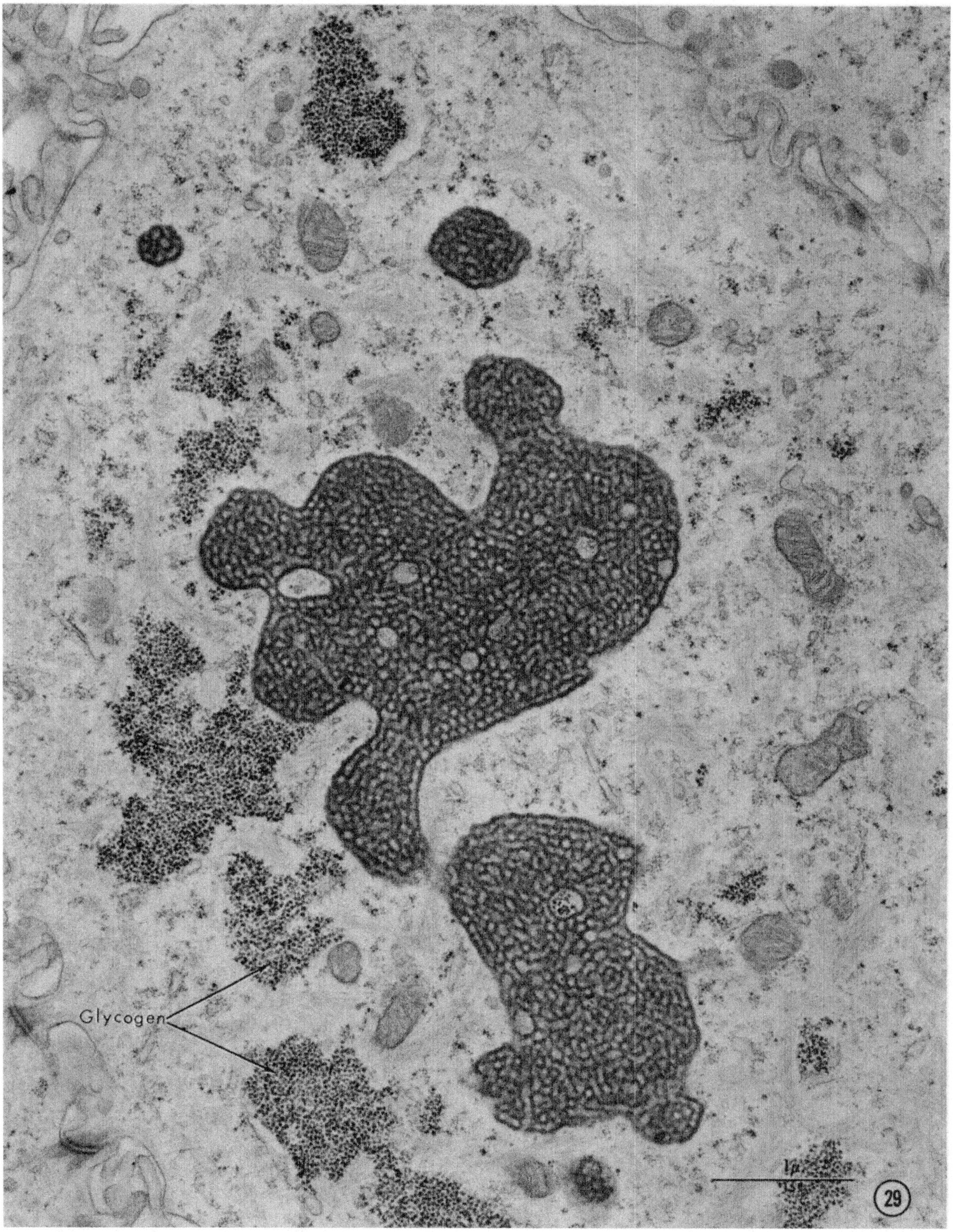

Fig. 29. Enlarged view of the mesh-like peridermal granules from a 16-day-old chick embryo. Veronal buffered osmium fixation. ×24,400.

B. Feather

More than by any other single feature, birds are characterized by feathers. Despite their light weight, feathers provide a most durable covering. Overlapping feathers entrap air and thus provide excellent insulation, reducing body heat loss and maintaining a constant high body temperature. Oily secretions from preening glands coat the feathers and repel water. Coloration, ranging from drab to dazzling, serves either to conceal and camouflage or to provide a brilliant display.

The shape and distribution of feather types are directly related to their aerodynamic function. Contour feathers, which are suited to flight, have a stiff central shaft with parallel branches called barbs. These barbs are attached to their neighbors by means of interdigitating barbules, upon which are interlocking hooklets or barbicels. Flightless birds, such as ostriches, have fluffy feathers without barbicels.

Down feathers, which provide good insulation, are found on newly hatched birds and under the contour feathers of many adults, especially aquatic ones. They consist of a reduced shaft, the calamus, with many fluffy barbs. An embryonic down feather consists of a delicate circlet of barbs surrounding a central pulp cavity containing blood vessels. The entire structure is wrapped in a protective sheath covered with periderm (Fig. 30). As the chick dries out after hatching, the protective sheath ruptures, releasing the numerous barbs with their attached barbules as fluffy down. The square outlined in Fig. 30 indicates the portion of feather shown in Fig. 31.

Feather germs appear for the first time in $6\frac{1}{2}$-day-old embryos, and by day 14 the developing feathers give an x-ray diffraction pattern identical with those of the adult. In the developing feather, the central part or pulp consists of connective tissue elements. Surrounding the pulp are the developing barb cells which contain presumptive β-keratin (Fig. 31). Similar to the chick skin, periderm consisting of two layers of cells surrounds the forming feather.

Even though β-type keratin is found in both reptilian scales and bird feathers, their mode of formation differs in the two tissues. In scales, a filamentous matrix complex appears in membrane-bounded packets that increase in size and number and eventually fill the cells. In feathers, filaments occur singly or group into bundles that are oriented parallel to the long axis of the feather (Fig. 32). These filaments are 30 Å in diameter, the α filaments are 80 Å. The most predominant organelle in these cells is ribonucleoproteins. The filaments increase rapidly in number, occupying the whole cell to the exclusion of the nucleus and other organelles.

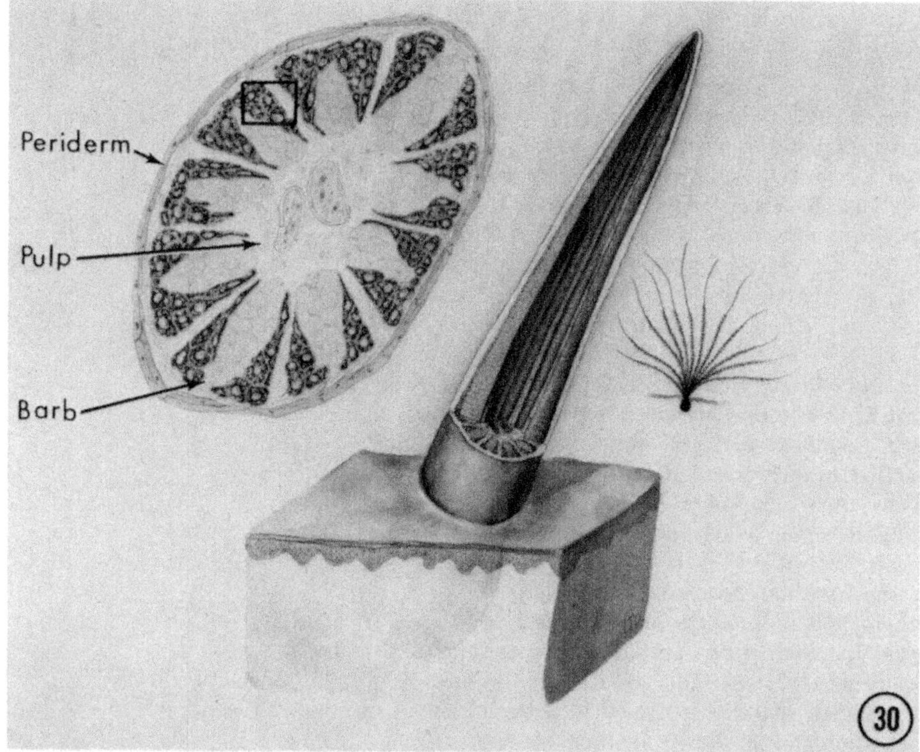

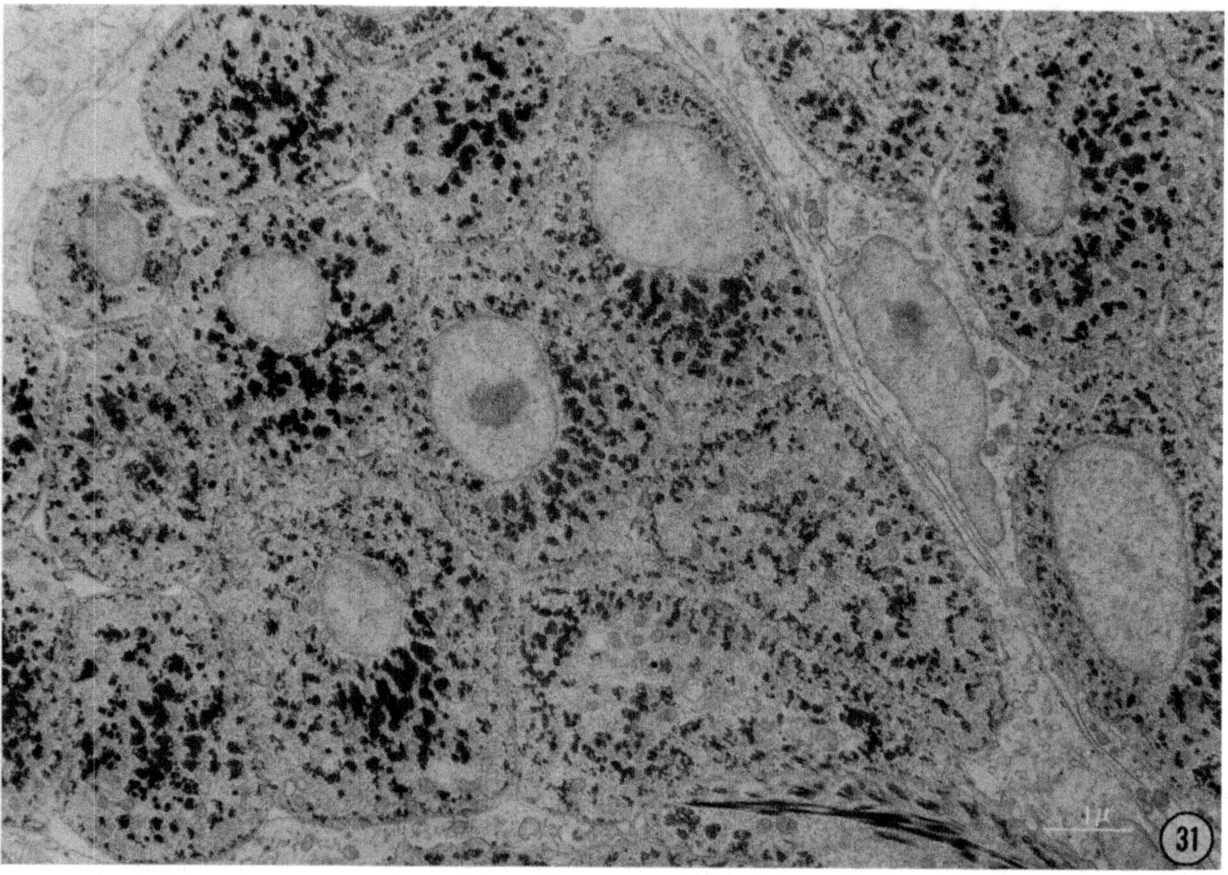

Fig. 30. Diagram of a forming down feather (after Rawles, 1965). Electron micrograph of a corresponding area in the box is shown in Fig. 31.

Fig. 31. Forming feather of a 14-day chick embryo. Veronal acetate buffered osmium fixation. ×14,400.

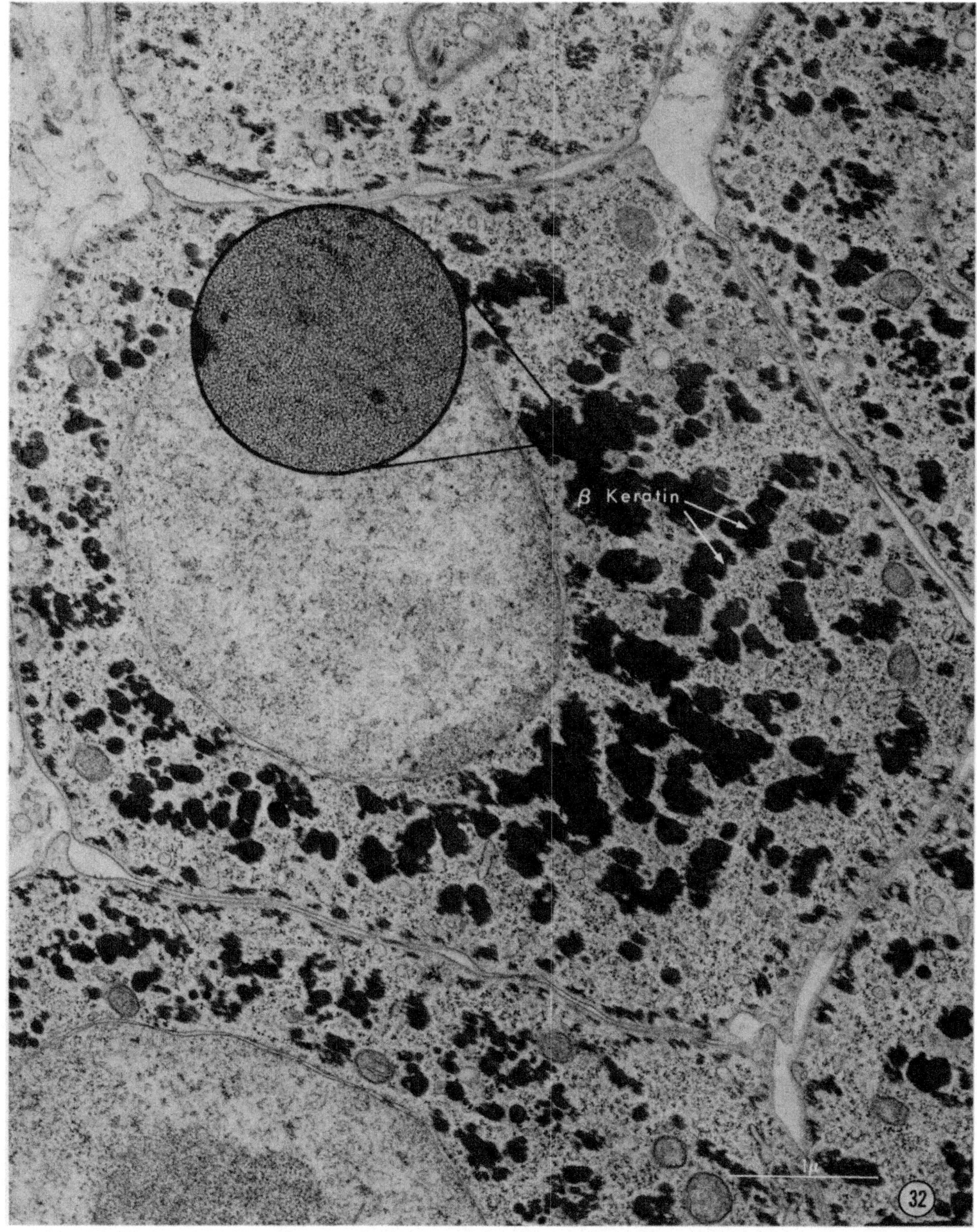

Fig. 32. Enlarged view of forming feather cells showing β-keratin formation. Inset shows tightly packed 30 Å filaments similar to those of reptilian β layer shown in Fig. 15. Veronal acetate buffered osmium fixation. ×26,000.

VI. MAMMALIAN EPIDERMIS

More than any other group of vertebrates, mammals exhibit extreme diversity in such keratinized structures as hairs, horns, hoof, nails, and scales. Mammalian epidermis varies widely in thickness and degree of keratinization from species to species and in different parts of the body of the same animal. The thick epidermis of nonhairy animals like the elephant and hippopotamus contrasts with the thin epidermis of fur bearers. In friction areas such as soles, palms, and prehensile tails, the epidermis is extremely thick. The stratified epithelium of mammalian epidermis also characterizes many of the internal cavities such as the oral mucosa, vagina, and conjunctiva.

Mammalian epidermis has four distinct layers: basal, spinous, granular, and horny (Figs. 33 and 34). The basal cells divide and during their upward migration undergo differentiation. During maturation, the cells increase in size and the Golgi complex and rough-surfaced endoplasmic reticulum become prominent. In

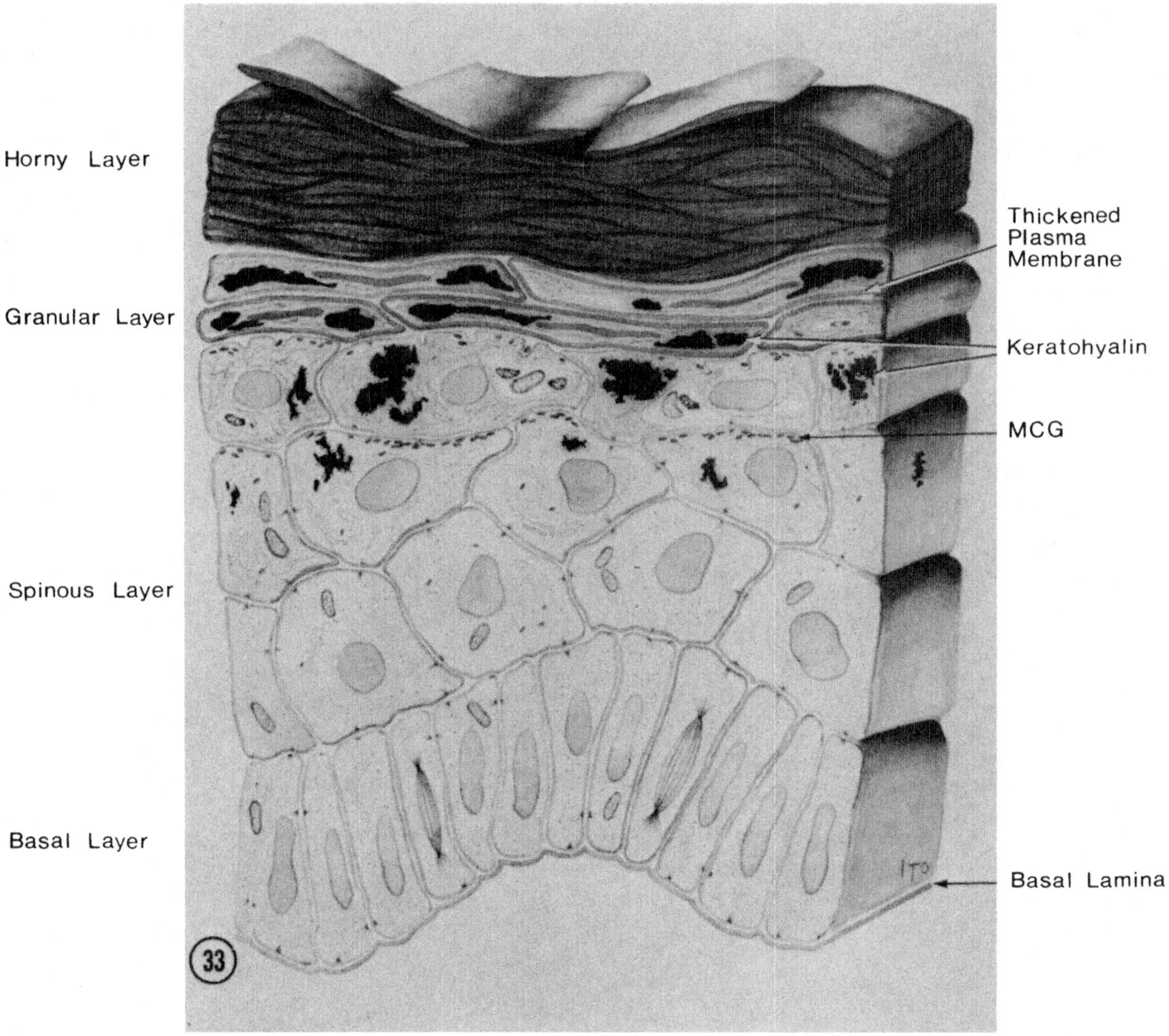

Fig. 33. Diagram of mammalian epidermis.

Fig. 34. Epidermis of the rhesus monkey (*Macaca mulatta*) showing the basal, spinous, granular, and horny layers. Collidine buffered glutaraldehyde. ×9000.

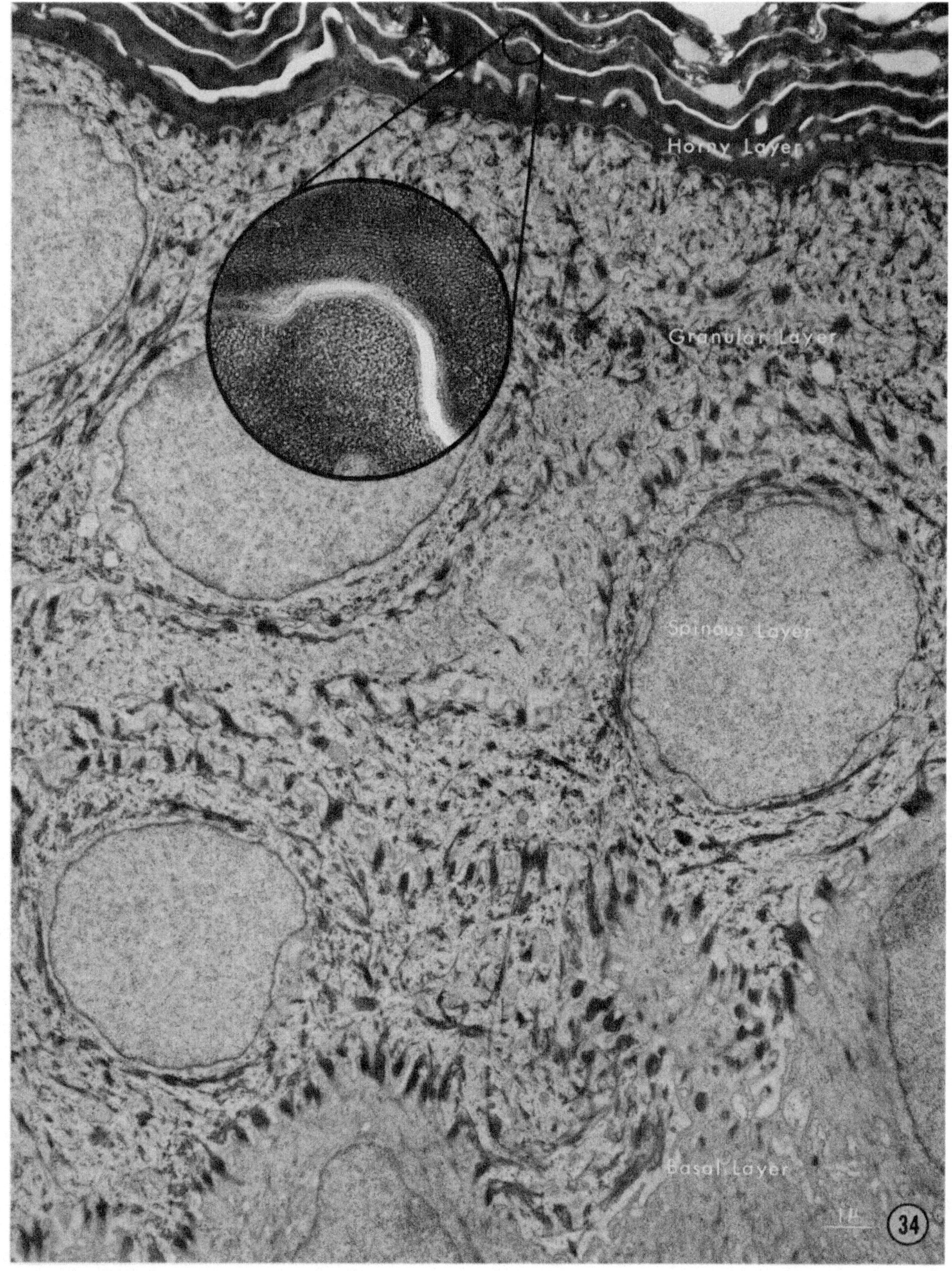
Horny Layer
Granular Layer
Spinous Layer
Basal Layer
1μ
34

close association with the Golgi vesicles, many membrane-coating granules (MCG; Odland bodies, or keratinosomes) are formed (Fig. 35). They are ovoid lamellated bodies about 0.2 μ in diameter. Each of the granules is membrane bounded and closely packed with thick and thin lamellae about 30 Å. Once formed, they migrate toward the cell periphery and are eventually discharged into the intercellular spaces, where they disintegrate and spread their contents over the plasma membranes (Figs. 36, 37, 38). The function of these granules remains controversial. Since they contain hydrolytic enzymes both before and after discharge, they may play a role in the exfoliation of the horny cells. Another theory is that the substance of the granules functions like a cement, binding the cells together and acting as a water barrier.

Fig. 35. Forming granules are seen in association with the Golgi complex in the spinous cells. ×44,200.

Fig. 36. Alignment of membrane-coating granules near the plasma membrane in the granular layer. ×50,600.

Fig. 37. After the release of the granules into the intercellular spaces many of them show the lamellar arrangement. ×40,000.

Fig. 38. Thick and thin plasma membranes at the junction between the horny and granular cells. Note the disintegrating membrane-coating granules. ×52,500.

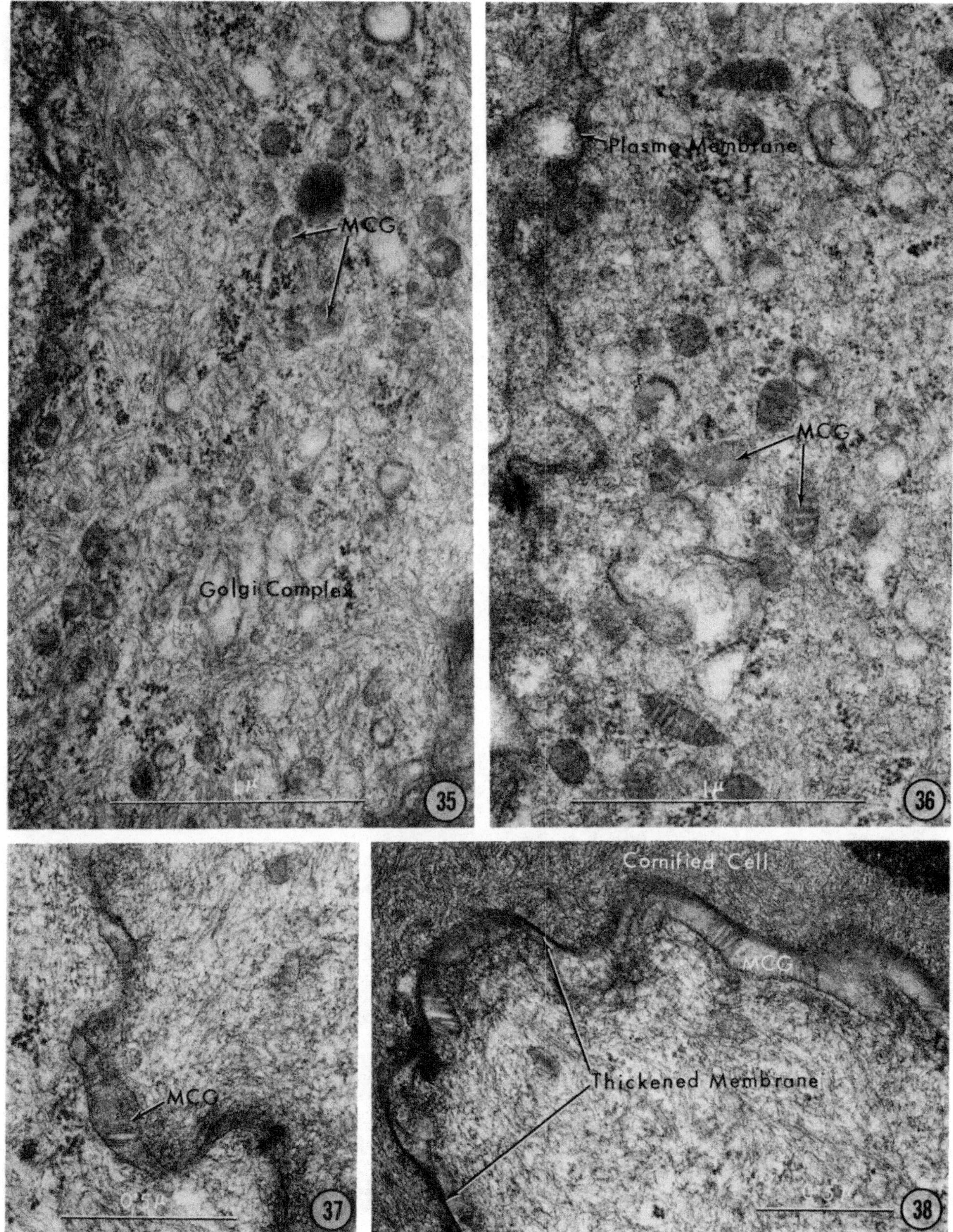
MCG
Golgi Complex
35
Plasma Membrane
MCG
36
MCG
37
Cornified Cell
MCG
Thickened Membrane
38

During differentiation, the spinous cells also form small amorphous keratohyalin granules that increase in size and number in the granular layer. They are polymorphic, irregularly shaped in man (Fig. 40) and spherically shaped in rodents (Fig. 41). Even though the main mass appears amorphous, bundles of filaments or clusters of ribosomes often appear in the immediate vicinity of the keratohyalin granules. Many substances, such as chromatin, derived from the breakdown of the nucleus, RNA, mucopolysaccharides, and lipids are reportedly present in keratohyalin granules. Recent autoradiographic studies have shown a histidine-rich protein in these granules. When the horny cells are formed, keratohyalin granules mysteriously disappear; their function is still a matter of speculation.

Another most important feature of the keratinized cell is the thickened plasma membrane (Figs. 38 and 39). In basal, spinous, and granular cells they are only about 80 Å thick, but in horny cells they range from 150 to 200 Å. The intracytoplasmic deposition of a band of material along the inner leaflet is thought to cause the membrane to thicken.

The horny cells are filled with 80 Å filaments, which are synthesized with differentiating cells and increase in number, eventually filling the cell. When packed tightly, these electron-lucent filaments are surrounded by a dense matrix and do not pick up stain. However, when not packed tightly as in the keratinized cells of the esophagus and oral mucosa, the filaments are electron dense and appear discrete.

Fig. 39. Horny layers of the esophageal epithelium of the mouse. Note that the filaments are not tightly packed and hence show no "keratin pattern." Veronal acetate buffered osmium fixation. ×28,000.

Figs. 40 and 41. Polymorphic keratohyalin granules. Figure 40 shows an irregularly shaped keratohyalin granule of human epidermis. ×44,000. Figure 41 shows oval shaped keratohyalin granules from rat oral mucosa surrounded by ribosome-like particles. ×53,000. In both the granules have electron-dense and electron-lucent areas. Veronal acetate buffered osmium fixation.

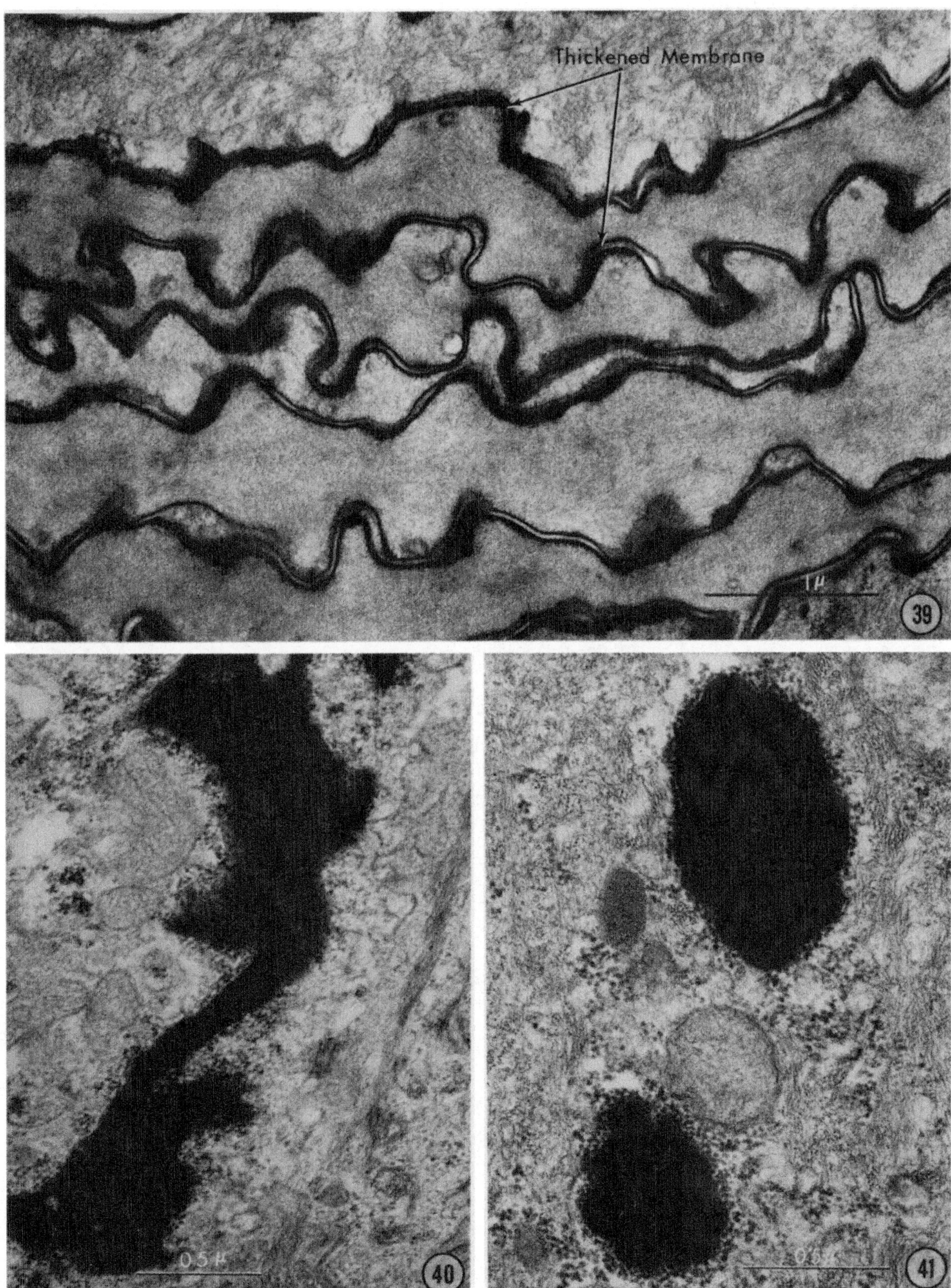
Thickened Membrane
1 μ
39
0.5 μ
40
41

A. Mammalian Fetal Epidermis

Early in embryonic development, the epidermis consists of only a single layer of germinative cells. As additional cell layers are formed, the superficial layer becomes the periderm, and the underlying layers the epidermis. The periderm provides protection for the developing epidermis and also helps in the exchange of substances between the uterine environment and the fetus.

The covering epithelium of a 79-day-old rhesus fetus (gestation period, 165 days) consists of a distinct layer of germinative cells overlaid by several layers of partially differentiated cells and a superficial layer of periderm. In several ways the development of mammalian skin is different from that of chick skin (see Section V,A and Figs. 27, 28, and 29). For example, mammalian periderm lacks the peridermal granules and large amounts of glycogen characteristic of birds. By contrast, cells of the mammalian presumptive epidermis have large accumulations of glycogen that are absent both in chick and adult mammalian epidermis. In the 79-day-old presumptive epidermis, the adult characteristics have not been established (Fig. 42). Horny or granular layers are lacking, the large cuboidal germinative cells rest upon a basal lamina, numerous intercellular spaces are evident between cells, and desmosomes are few. As the cells differentiate, desmosomes increase. The 80 Å-thick filaments characteristic of all stratified epithelia are aligned along the cell periphery.

At about 50 days gestation, the basal cells divide and grow downward, giving rise to the anlage of skin appendages such as hair and sweat glands. By 150 days, both hair and sweat glands are fully developed.

Fig. 42. Fetal back skin from a 79-day-old rhesus monkey (*Macaca mulatta*) showing the superficial periderm and the presumptive epidermis underneath. Veronal acetate buffered osmium fixation. ×9500. (Micrograph courtesy of Dr. M. Bell.)

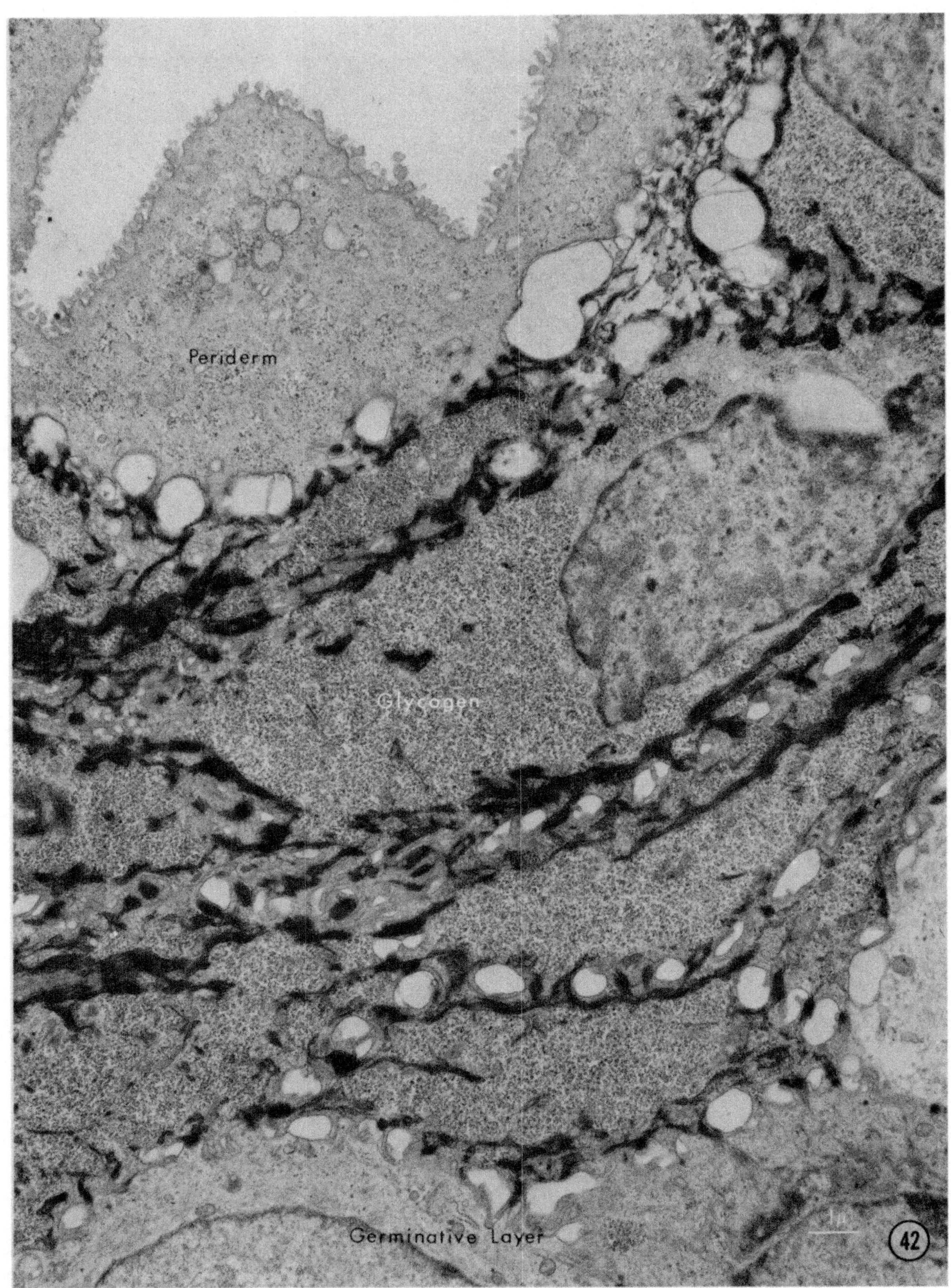
Periderm
Glycogen
Germinative Layer
42

B. Vaginal Epithelium

Vaginal epithelium, one of the target organs for ovarian hormones, undergoes keratinization because of the influence of estrogen. In spayed rhesus monkeys, the epithelium atrophies and consists of only two or three layers of undifferentiated cells. During the follicular phase when estrogen levels are high or after the administration of estrogen into spayed animals, the epithelium hypertrophies into forty or fifty layers of cells. The basal cells divide and the daughter cells keratinize as they migrate upward. In the spinous cells, many membrane-coating granules (MCG) appear which later migrate toward the apical cell surface where they are discharged into the intercellular spaces (Fig. 43). Keratohyalin granules are also formed in the cells; however, they are smaller and fewer than in the epidermis. The most characteristic product of the vaginal epithelium is glycogen, synthesized in large quantities in the differentiating cells (Fig. 44). Glycogen content increases rapidly, almost completely filling the superficial cells (Fig. 43).

Even the exfoliated cells have large amounts of glycogen which is fermented by bacteria to provide an acidic environment for the vaginal lumen. Unlike those in the epidermis, the exfoliating vaginal cells have nuclei and many of the cytoplasmic organelles.

Fig. 43. Large quantities of glycogen in the superficial cells of the vaginal epithelium of the rhesus monkey (*Macaca mulatta*). Veronal acetate buffered osmium fixation. ×20,700.

Fig. 44. Small accumulations of glycogen in the basal cell of the rhesus monkey vaginal epithelium. Veronal acetate buffered osmium fixation. ×14,400.

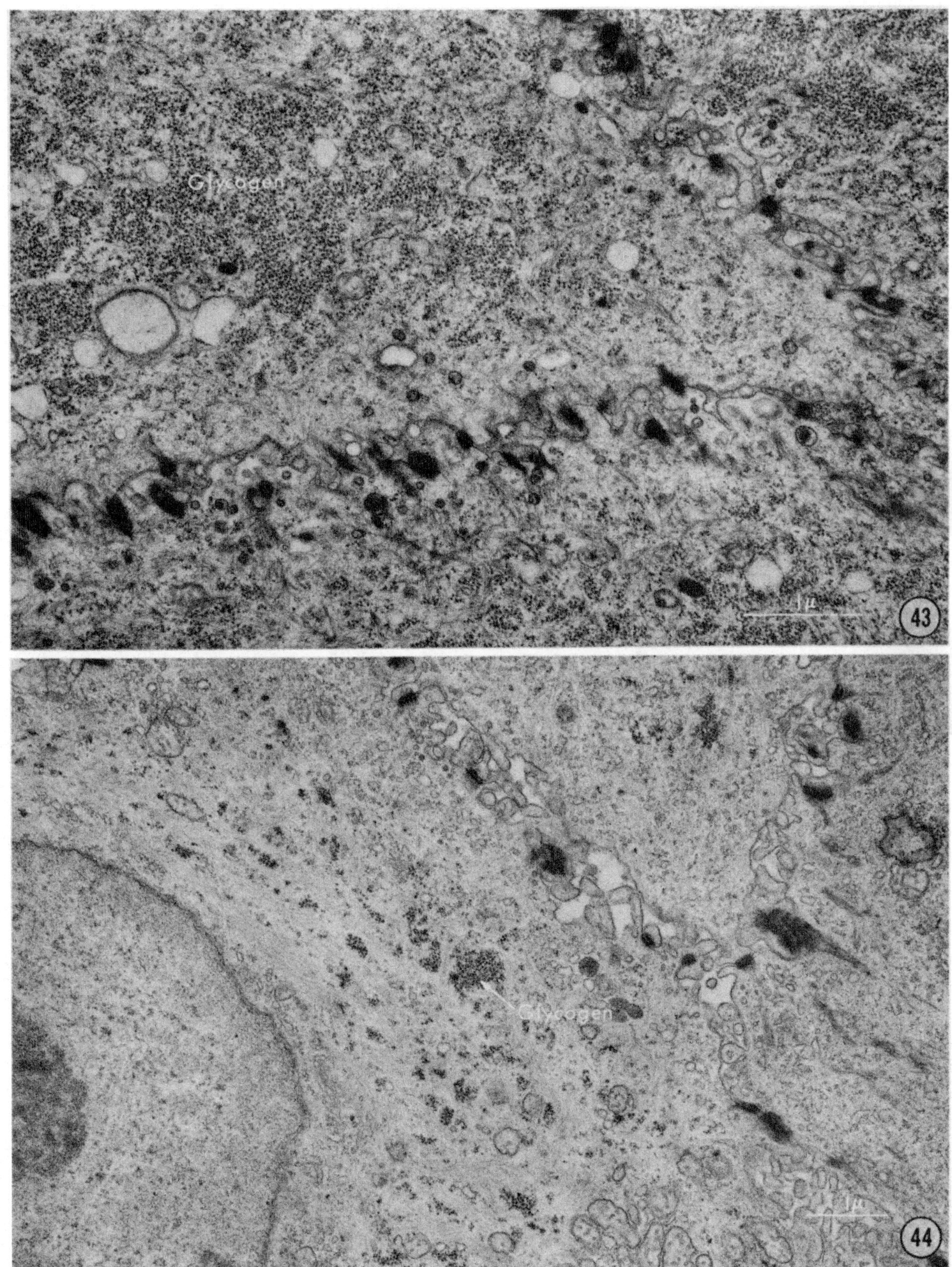

C. Conjunctival Epithelium

The epidermis of the eyelid is continuous with the conjunctiva or inner surface of the eyelid. Whereas the epidermis is characterized by keratinizing stratified epithelia, the conjunctival epithelium produces mucus. At the junction between these two epithelia, the characteristics of both merge (Fig. 45). As the epidermis approaches the conjunctiva, the cornified layer becomes thinner until at the junction the superficial cells are not cornified. In the same way, the granular layer becomes less pronounced until in the cells immediately below the superficial cells only a few small keratohyalin granules remain. Membrane-coating granules are aligned at the apical cell surface as in keratinizing epithelia.

In the conjunctiva proper, the superficial cells are nucleated and contain all the organelles and many filaments (Fig. 46). Many wide intercellular spaces are present between the superficial cells and the cells below, these two layers being adherent only at desmosomal contact points. The apical surface of the superficial cells has many microvilli with a coating of fuzz. The most characteristic product of the differentiating cells of the conjunctiva is mucous granules, which presumably are synthesized in the several Golgi areas distributed in the cells (Fig. 47). The mature granules, about 1500 Å in diameter, are round or oval and filled with a finely granular material enclosed in a smooth-surfaced membrane. Many of the granules migrate toward the apical plasma membrane where they either are secreted into the intercellular space or become dispersed in the cytoplasm (Fig. 47).

As the conjunctiva approaches the fornix of the eyeball, many mucous cells are distributed in the epithelium. They look like goblet cells filled with large mucous granules which are secreted to supply the lubricating mucus (Fig. 48).

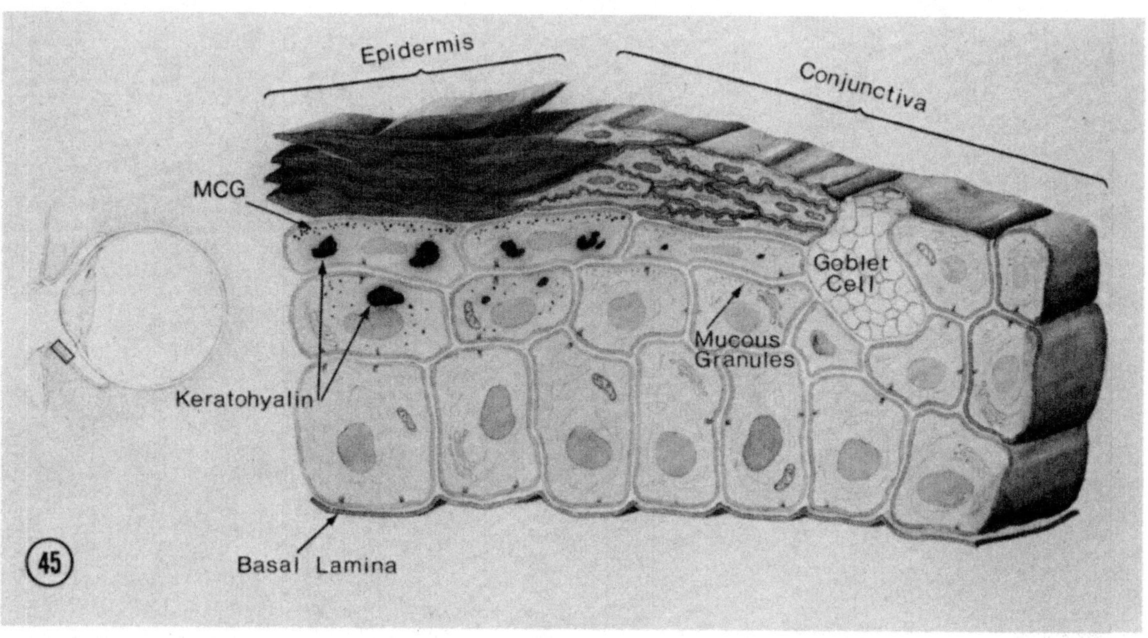

Fig. 45. Diagram of the junction between the epidermis of the eyelid and conjunctiva. Diagram corresponds to box indicated in left drawing.

Fig. 46. Transitional zone between the keratinizing and mucus-producing stratified epithelia of the mouse. Veronal acetate buffered osmium fixation. ×15,700.

Fig. 47. Superficial cells of the mucus-producing conjunctiva. Veronal acetate buffered osmium fixation. ×14,000.

Fig. 48. Goblet cells of the conjunctiva. Veronal acetate buffered osmium fixation. ×16,000.

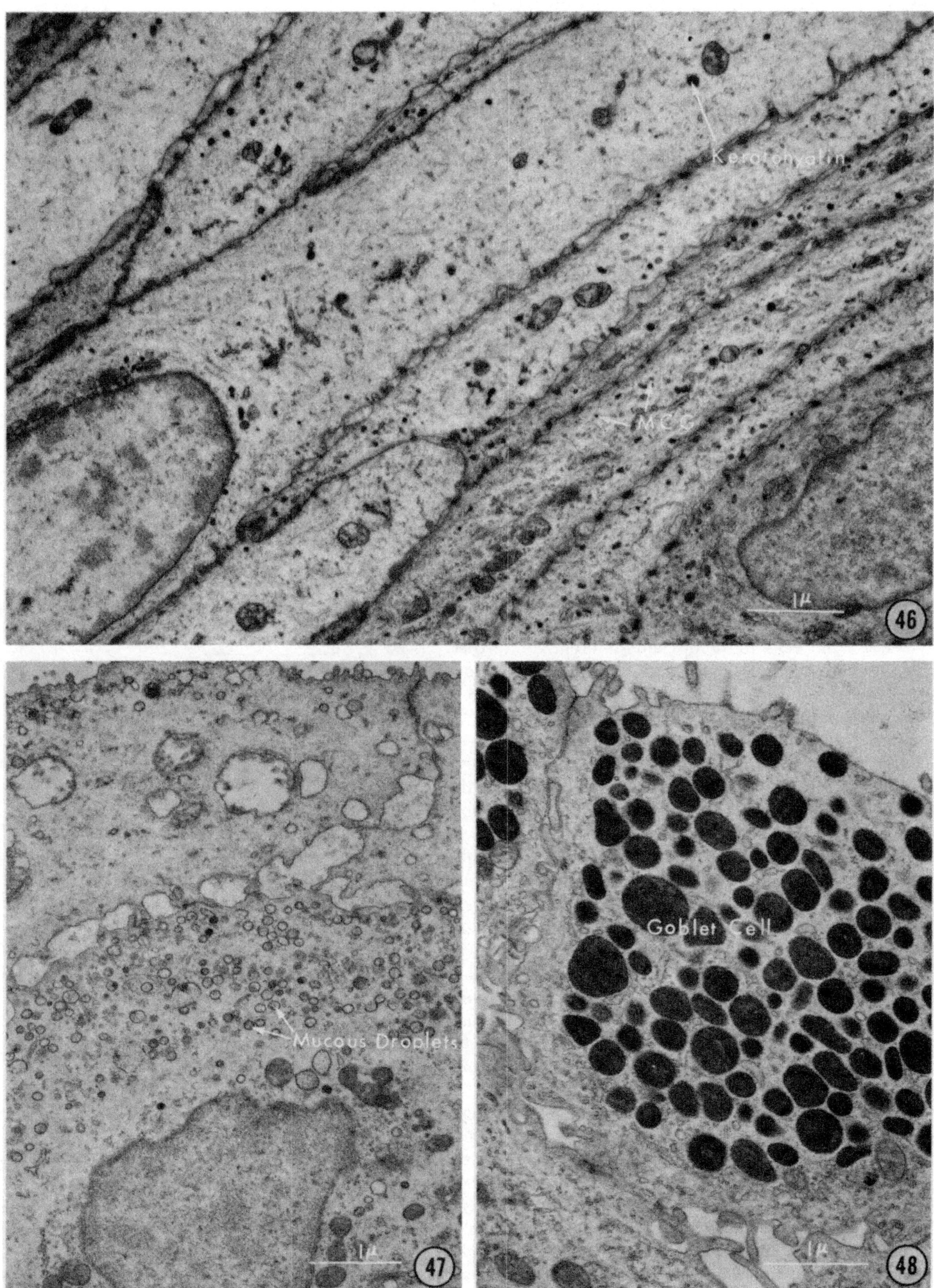
Keratohyalin
MCG
46
Mucous Droplets
47
Goblet Cell
48

D. Anagen Hair Follicle

The hair growth cycle is one of the lesser known of the many biological cycles. During the cycle, mammalian hair passes from active growth (anagen) through an intermediate phase (catagen) to a quiescent state (telogen). The anagen follicle extends deep into the dermis where it expands into a bulb composed of mitotically active matrix cells that produce the emerging hair. After a period of growth, which can last several years, the follicle enters into catagen, when it is reorganized into the resting follicle. The telogen follicle has ceased to grow but by means of a club, holds tightly to the hair that was produced during the preceding anagen phase.

The anagen hair follicle, which produces hair continuously, reflects the complex morphogenetic movements of the different layers of the follicle (Fig. 49). The matrix cells of the bulb divide rapidly; as the daughter cells move upwards, they differentiate into the hair shaft (medulla, cortex, and cuticle) and the three layers of the inner root sheath (Fig. 50). The different cell layers can be identified by their location in the hair follicle as well as by their characteristic structures and products (Figs. 50 and 51). During the early stages of differentiation, the centrally located medullary cells develop amorphous granules of various sizes. As the cells move up and differentiate further, the granules continue to grow, reaching several microns in diameter until they fill a large portion of the cell. Simultaneously the medullary cells develop a moderate number of membrane-limited vesicles. At advanced stages, the nucleus and cytoplasmic organelles begin to disintegrate. The cells around the medulla form the cortex of the hair. During the process of maturation, the cortical cells synthesize filaments about 80 Å in diameter which aggregate into bundles that eventually fill the cells. In the completely keratinized cortical cells, the circular profiles of the filaments are electron-lucent and are surrounded by an electron-dense matrix, an arrangement known as the "keratin pattern." The cuticle forms the outermost layer of the emerging hair. During maturation the cuticular cells first form granules about 300–400 Å, which grow to 5000 Å. Concomitantly they migrate toward the cell membrane. Eventually the keratinized, flattened, and overlapping cuticular cells are filled with coalesced masses of granules.

The layers of the inner root sheath (cuticle, Huxley's and Henle's layer) develop in the same way. Early in the differentiation, the cells develop numerous trichohyalin granules that appear in amorphous masses, often with filaments protruding from them (Figs. 50 and 51). These granules grow until they fill the cells; fully keratinized inner root sheath cells are replete with filaments. The keratinized innermost sheath cells are shed into the pilosebaceous canal. Surrounding the inner root sheath is the multilayered outer root sheath, continuous with the epidermis at the emerging end of the hair follicle but extremely attenuated at the lower region of the bulb. The cells of the outer root sheath have large accumulations of glycogen.

The connective tissue elements of the hair follicle consist of the dermal papilla enclosed by the bulb and the connective tissue sheath surrounding the hair follicle. This latter element consists of a basal lamina and two layers of orthogonally arranged collagen fibers. This, in turn, is surrounded by a loosely arrayed layer of fibroblasts and macrophages, the cellular part of the connective tissue sheath, which is continuous with the papillary layer of the dermis.

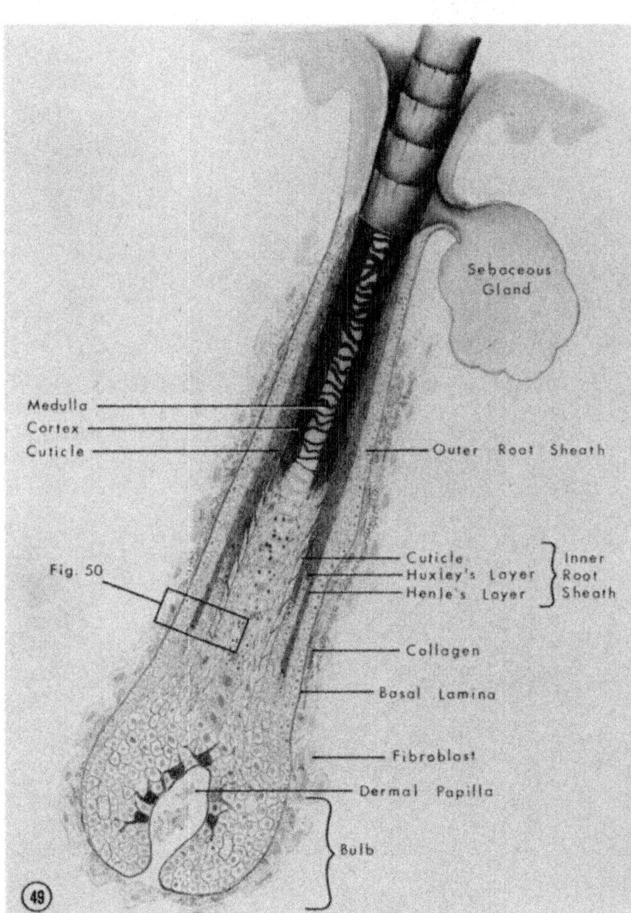

Fig. 49. Diagram of anagen follicle.

Fig. 50. Longitudinal section of the anagen mouse hair follicle showing complex arrangement of layers: medulla, cortex, cuticle, inner root sheath (I.R.S.), and outer root sheath. Veronal acetate buffered osmium fixation. ×10,300.

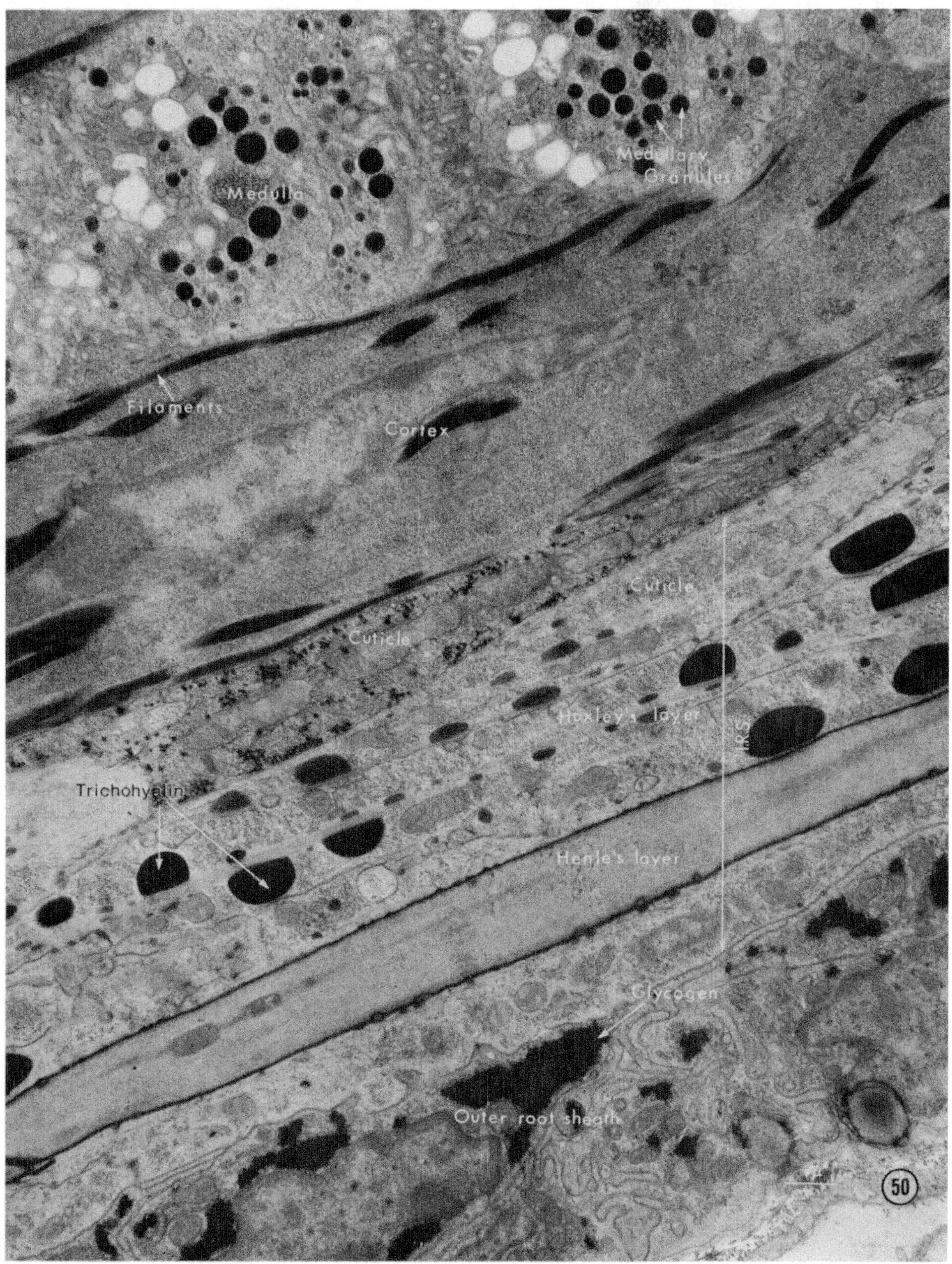
Medulla
Medullary Granules
Filaments
Cortex
Cuticle
Cuticle
Huxley's layer
IRS
Trichohyalin
Henle's layer
Glycogen
Outer root sheath
50

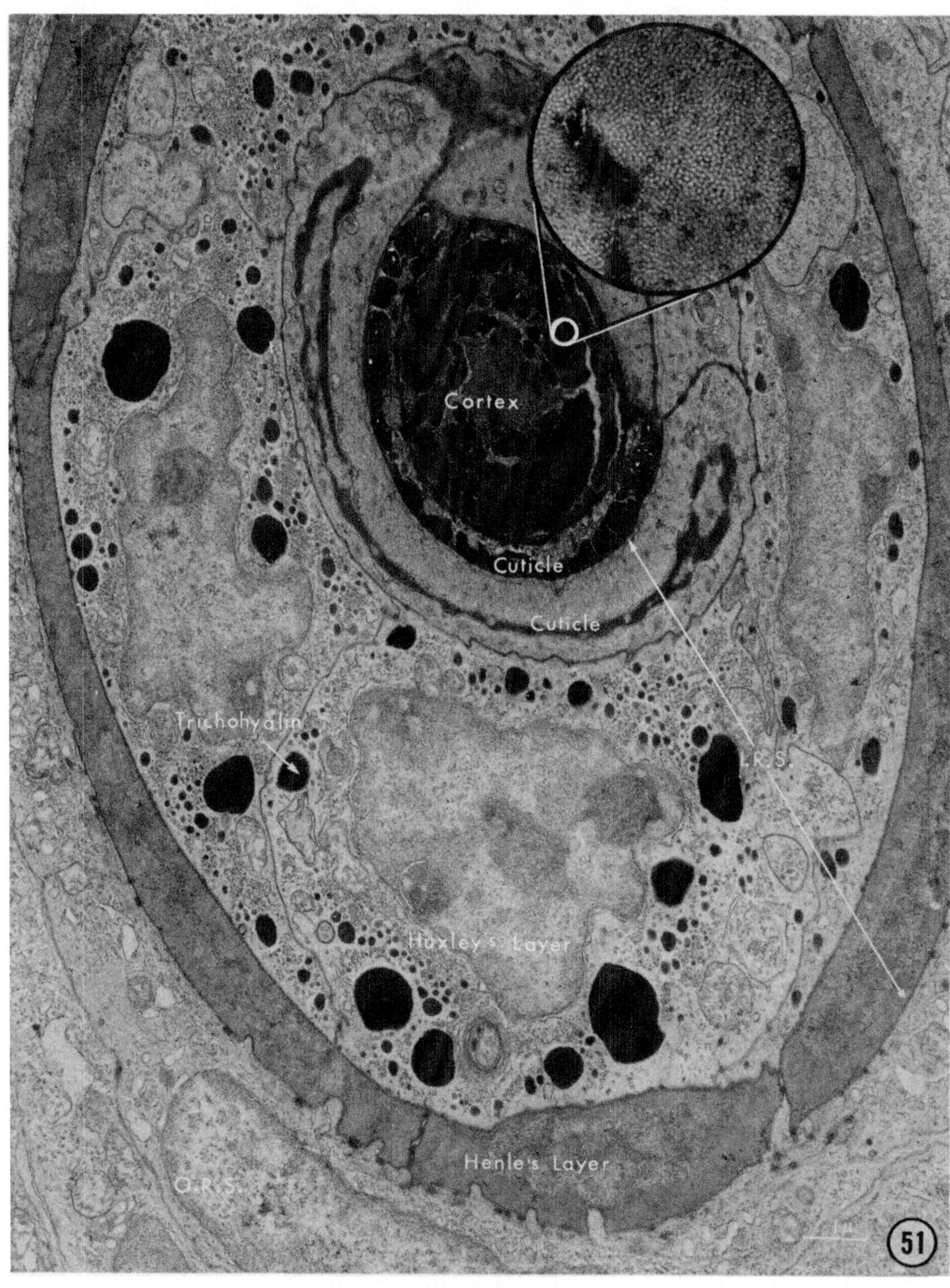

Fig. 51. Cross section of the anagen mouse hair follicle showing the different layers. Veronal acetate buffered osmium fixation. ×10,500.

E. Catagen Hair Follicle

After the growing (anagen) follicle has entered catagen, many of the structures of the growing follicles are eliminated as new structures of the resting follicle are formed. The onset of catagen is heralded by a cessation of mitotic activity in the matrix cells. However, cells that are already partly differentiated continue to differentiate and migrate upward to form the last part of the hair shaft.

This part is attached to the newly formed club, consisting of modified cortical cells filled with nonoriented filaments. Surrounding this club are the germ cells, which are formed by the transformation of the outer root sheath cells during catagen. Once the club and the surrounding germ cells are formed, the follicle below them is completely resorbed (Fig. 52). The epithelial cells are resorbed by cellular autophagy, the collagen fibers around the follicle by heterophagy.

As a prelude to this resorption, autophagic vacuoles begin to appear in the epithelial cells (Figs. 52 and 53). They grow in size and number and contain large amounts of acid phosphatase. Many of the vacuoles contain recognizable mitochondria, ribosomes, and glycogen. As the resorption continues, the cells are completely disintegrated. As more and more of the epithelial cells disintegrate, the basal lamina surrounding the lower portion of the follicle loses contact with the dying cells and begins to fold. In the later stages of catagen, the highly pleated basal lamina encloses remnants of the atrophic epithelial cells (Fig. 54). It is this folded basal lamina that appears as the wrinkled and thickened hyaline membrane of the catagen follicle of light microscopy. How the basal lamina is completely resorbed at the end of catagen is still obscure.

The layers of collagen fibers around the follicle are also broken down during this time. In many of the remodeling systems, collagenase, which is active at neutral pH and physiological temperatures, is responsible for the initial attack on the collagen molecule. After this partial breakdown, collagen is phagocytosed by macrophages and completely broken down by hydrolytic enzymes (Fig. 54). During catagen the follicle is surrounded by macrophages which have in their cytoplasm many vacuoles containing collagen in various stages of disintegration. Apparently, macrophages selectively engulf the collagen around the hair follicle. This activity is reflected in the highly folded plasma membrane of the macrophages facing the hair (Fig. 54).

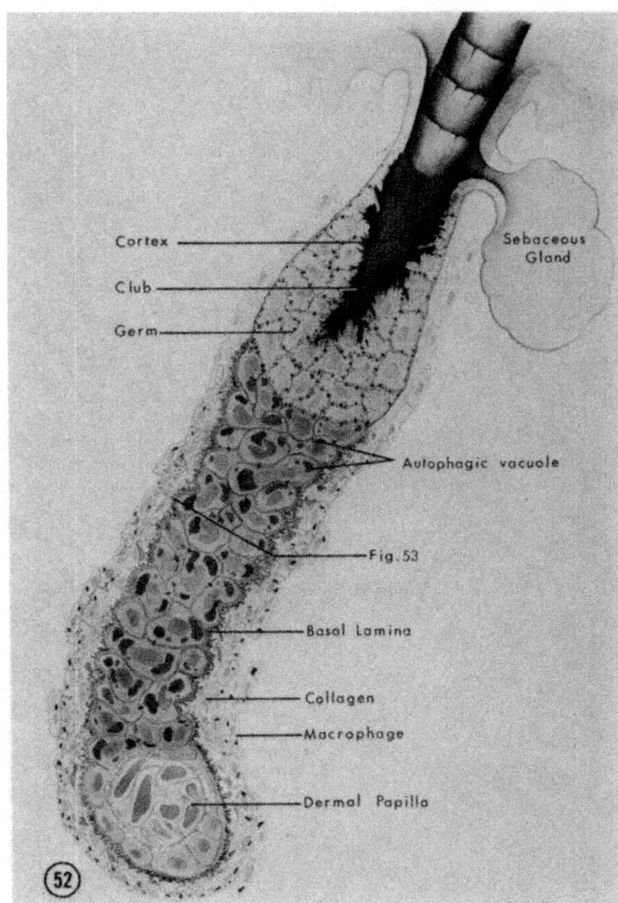

Fig. 52. Diagram of the catagen follicle.

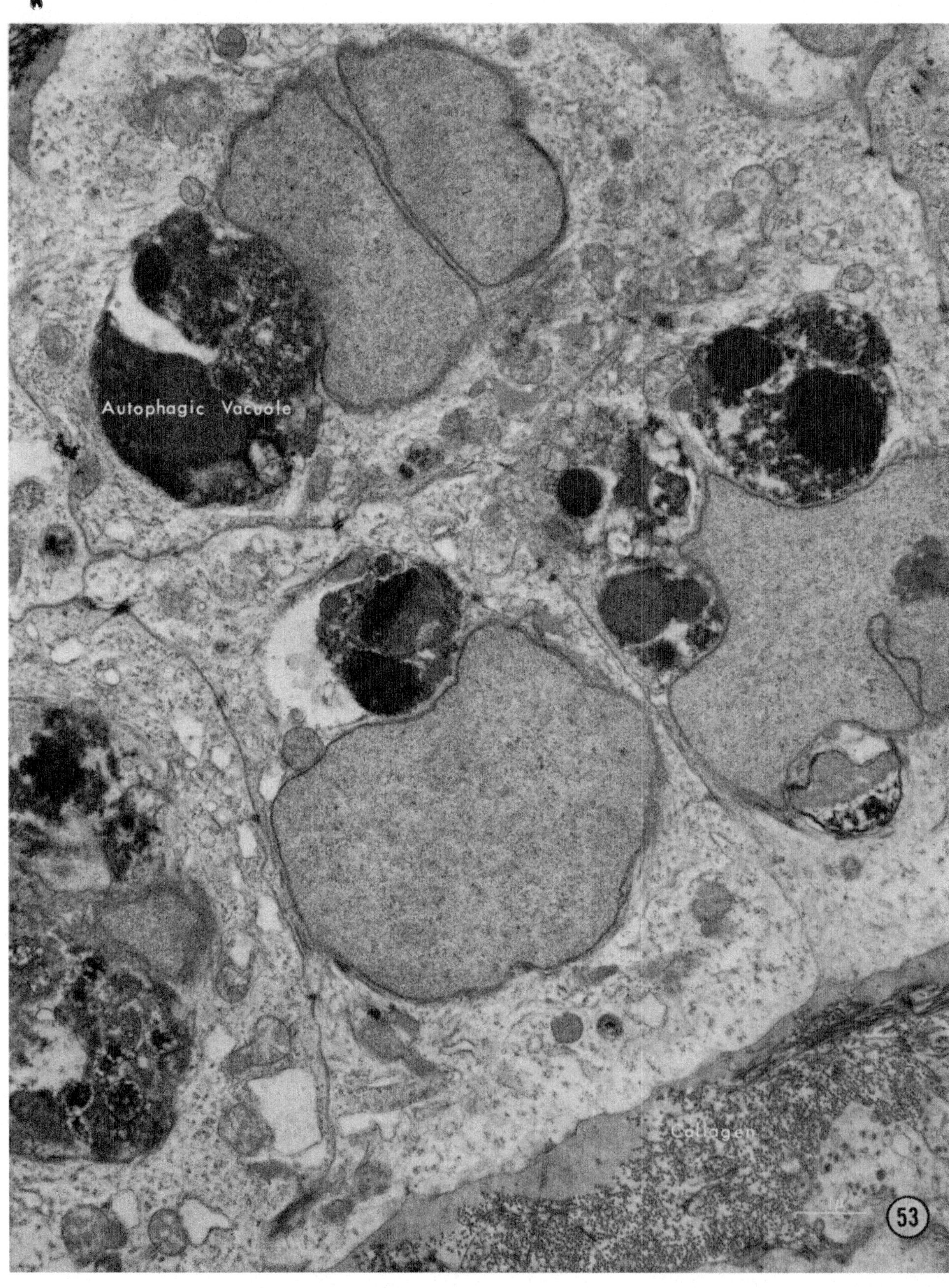

Fig. 53. Autophagic vacuoles of the cells undergoing resorption in the catagen follicle. Veronal acetate buffered osmium fixation. ×10,700. [From Parakkal, P. F. (1970). *Z. Zellforsch.* **107,** 81. Courtesy of the publisher.]

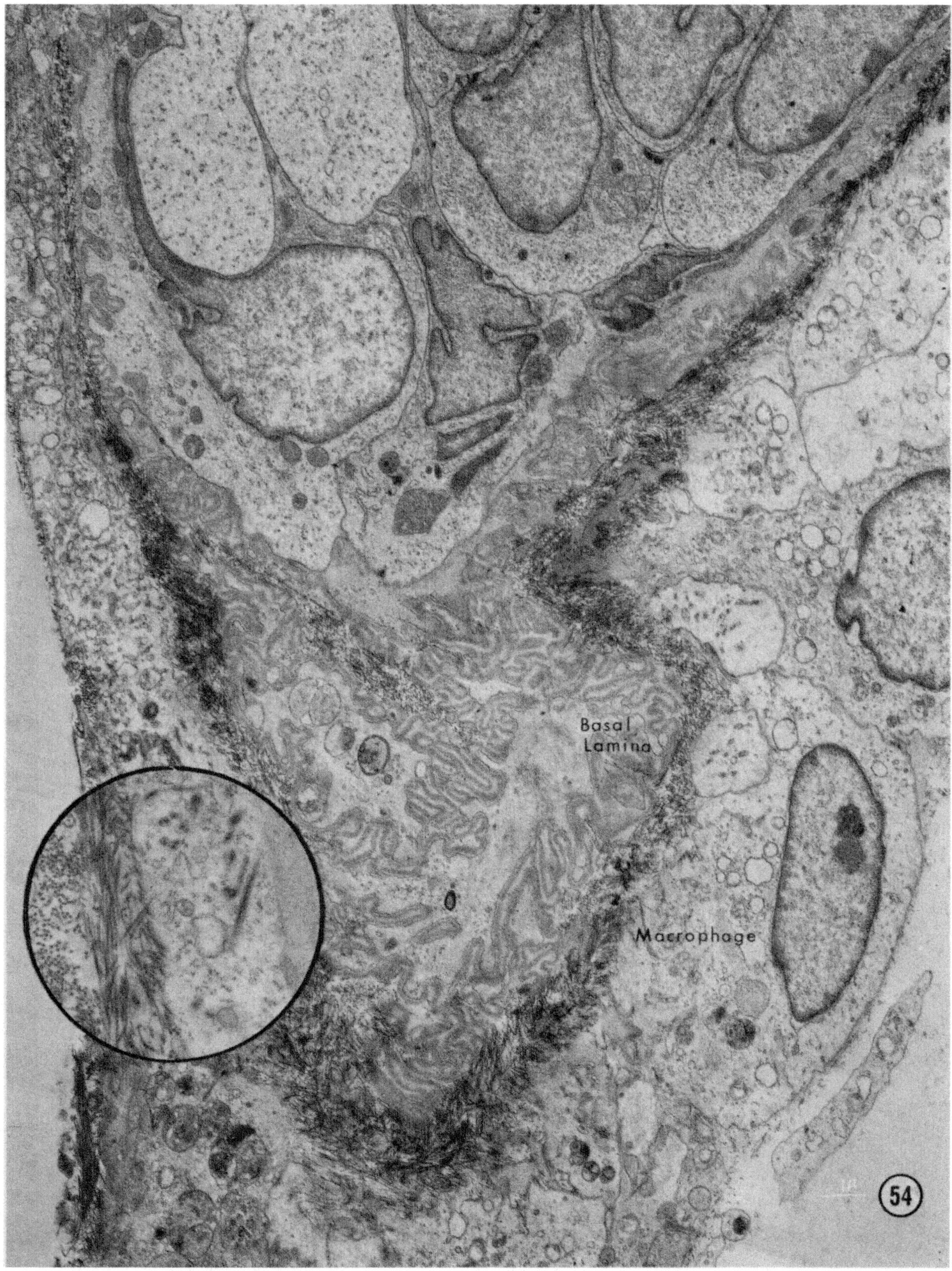

Fig. 54. Lower portion of the catagen follicle showing folded basal lamina surrounded by macrophages in the phagocytosed collagen. Inset: Portion of macrophage in the ingested collagen. Veronal acetate buffered osmium fixation. ×7,000.

F. Telogen Hair Follicle

One of the main functions of the telogen or resting follicle is to hold in place the hair produced during the preceding anagen. The telogen follicle can also regenerate the next generation of anagen hair. These dual functions are reflected in the structure of the telogen follicle. Many of the layers of the growing follicle, such as the medulla, cuticle, and inner root sheath are absent during the resting stage. The new structures that characterize the resting follicle are the club and the surrounding germ (Figs. 55 and 56). The club is attached to the hair shaft at one end and to the germ at the other. The club is responsible for anchoring the hair in position, and the germ cells give rise to the next generation of anagen hair. The club cells are modified cortical cells filled with disoriented 80 Å filaments (Figs. 56). The germ cells are produced by the transformation of the outer root sheath cells at the middle level of the anagen follicle. At the beginning of their formation, numerous autophagic vacuoles containing acid phosphatase activity begin to appear in the cells of the outer root sheath. These vacuoles also contain recognizable mitochondria, ribosomes, and large amounts of glycogen. As these organelles are resorbed, cytoplasmic filaments about 80 Å in diameter begin to form and eventually occupy large areas. By this time, the plasma membranes have developed numerous desmosomes.

The dermal papilla now looks like a ball of cells underneath the germ. These modified fibroblasts have nuclei filling the whole cells; scarcely any endoplasmic reticulum or Golgi apparatus, the characteristic organelles of active fibroblasts, remains.

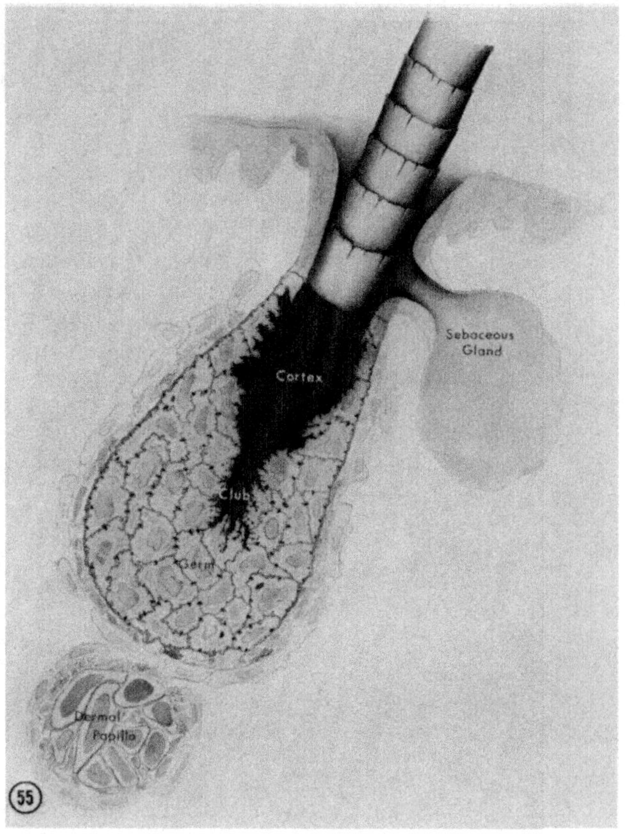

Fig. 55. Diagram of telogen follicle.

Fig. 56. View of the mouse telogen follicle showing the club cells interdigitating with the surrounding germ cells. Lower inset shows the junctional complex. Upper inset, longitudinal alignment of the 80 Å filaments of the cortex. Veronal acetate buffered osmium fixation. ×10,000.

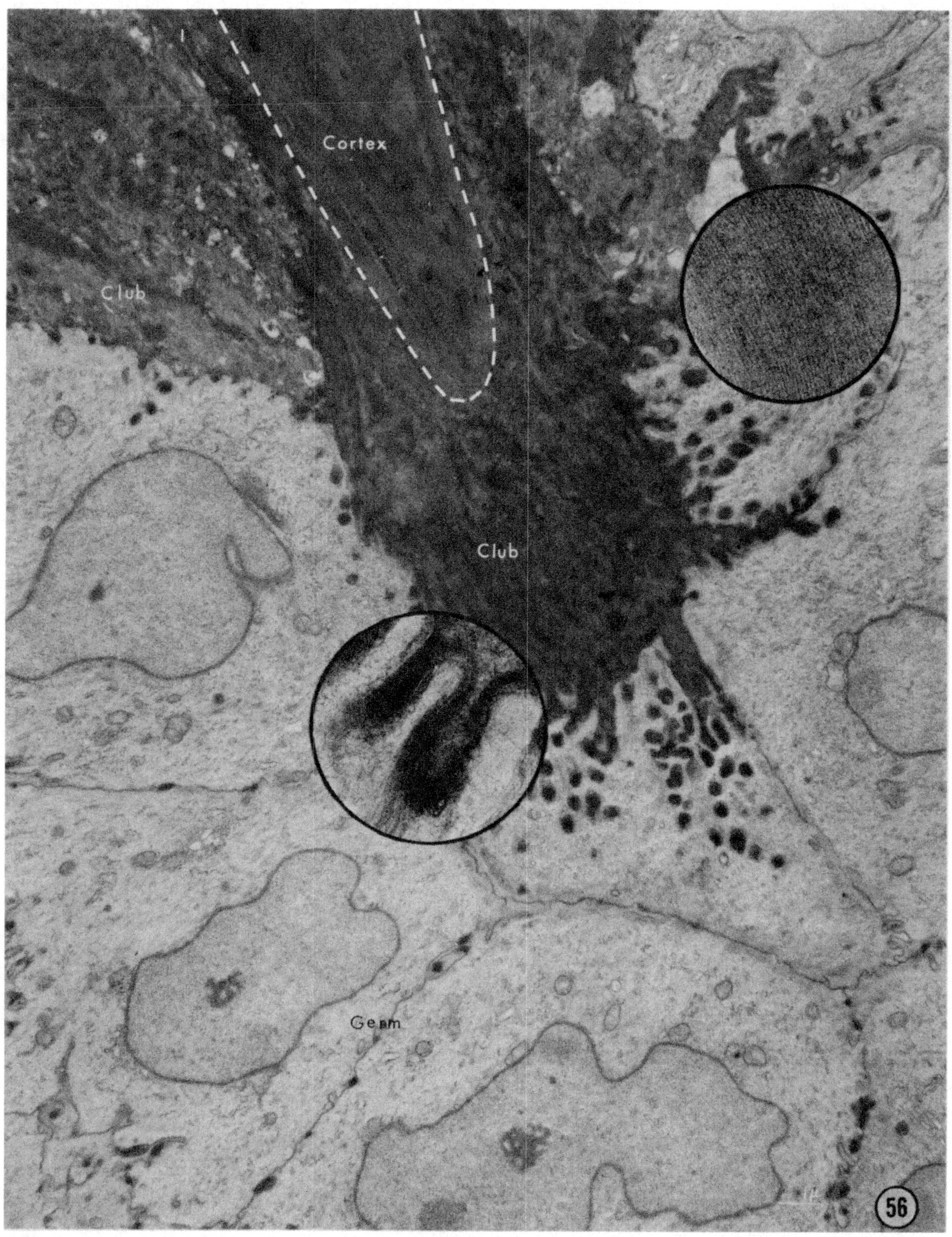
Cortex
Club
Club
Germ
56

G. Fingernail

Nails are tough keratinized structures tipping the digits of many primates and serving primarily as protective structures (Fig. 57). Nail is produced mainly by the cells of the matrix, which extends both dorsally and ventrally. The ventral matrix is larger than the dorsal and extends beyond the proximal nail fold. Through the nail surface, the ventral matrix appears whitish; it is the lunula of half-moon visible at the fingernail base. As the progeny of the matrix cells form the nail plate, it slides along the nail bed which contributes to the nail plate only to a very limited extent.

The basal matrix cells are similar to those of the epidermis in that besides the cytoplasmic organelles they contain many 80 Å filaments. These filaments, once formed, are organized into bundles which exhibit a "keratin pattern," namely, a tight packing of electron-lucent filaments surrounded by electron-dense matrix (Fig. 58). As the cells move upward and form the nail plate, they become completely filled with the filaments; simultaneously all of the organelles are dissolved. Even though keratohyalin granules are absent in the adult nail, they are present during the embryogenesis of the nail. The nail plate is composed of keratinized cells which are held together tightly and are not exfoliated singly like the epidermal cells.

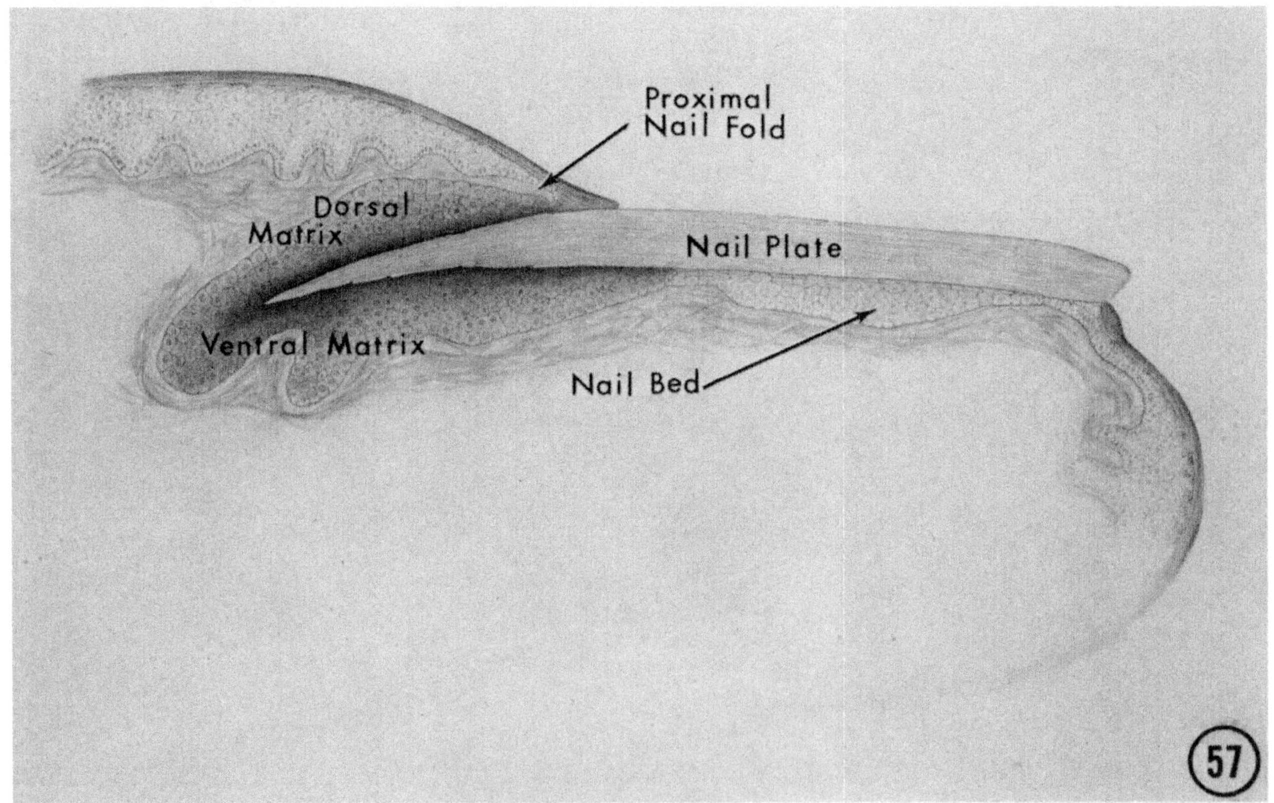

Fig. 57. Diagram of the fingernail.

Fig. 58. Junction between the nail plate and the underlying matrix cells from the fingernail of an infant rhesus monkey (*Macaca mulatta*). Right inset shows the "keratin pattern" of the nail plate. Left inset shows the forming "keratin pattern" in the matrix cells. Phosphate buffered glutaraldehyde, osmium postfixation. ×11,500.

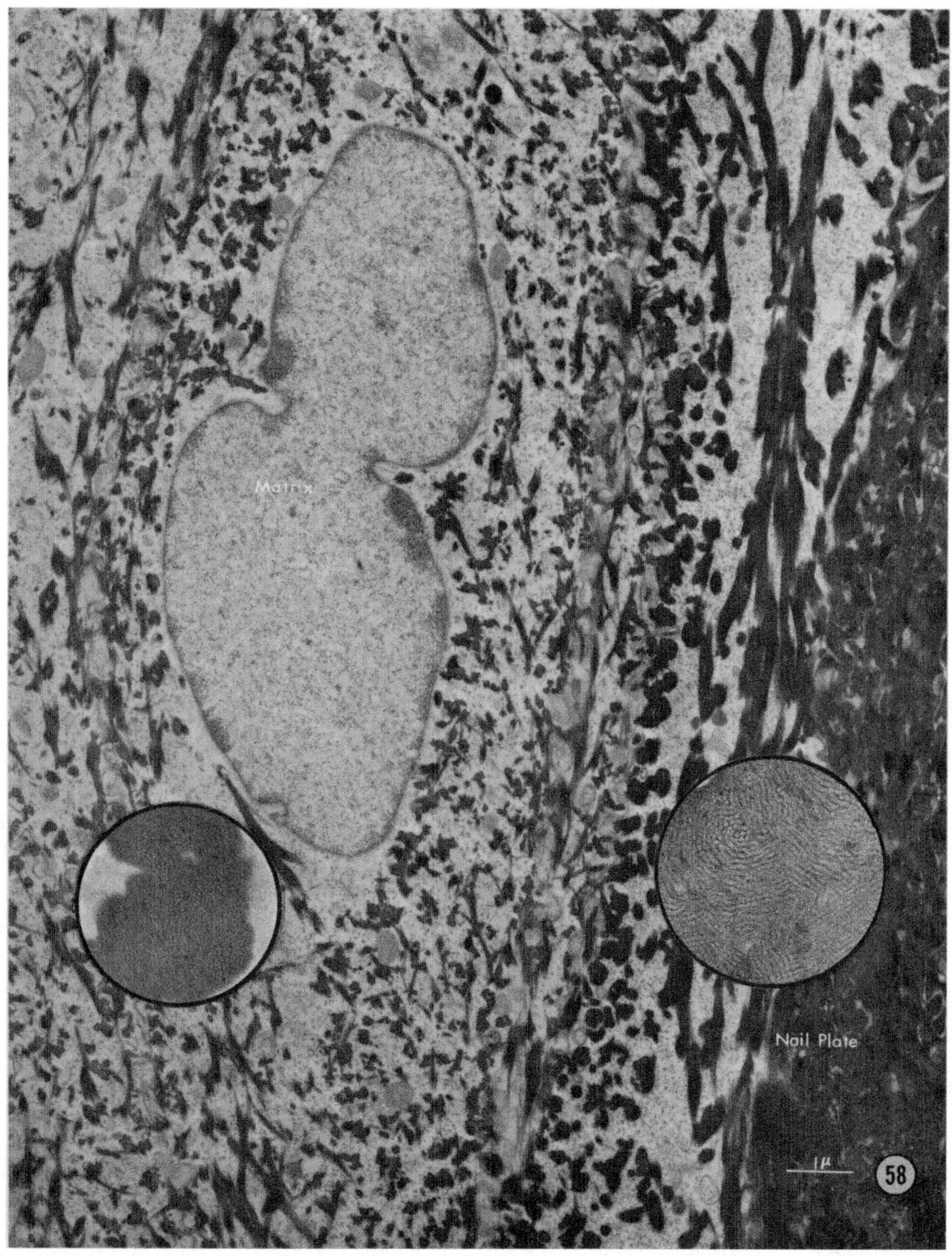

Matrix
Nail Plate
1μ
58

REFERENCES

1. Alexander, N. J. (1970). Comparison of α and β keratin in reptiles. *Z. Zellforsch.* **110**, 153–165.
2. Billingham, R. E., and Silvers, W. K., (1963). The origin and conservation of epidermal specificities. *N. Eng. J. Med.* **268**, 477–480, 539–545.
3. Brody, I. (1959). The keratinization of epidermal cells of normal guinea pig skin as revealed by electron microscopy. *J. Ultrastruct. Res.* **2**, 482–511.
4. Crick, F. H. C. (1953). The packing of α-helixes: simple coiled-coils. *Acta Crystallogr.* **6**, 689–697.
5. Downes, A. M., Sharry, L. F., and Rogers, G. E. (1963). Separate synthesis of fibrillar and matrix protein in the formation of keratin. *Nature (London)* **199**, 1059–1061.
6. Farbman, A. I. (1966). Plasma membrane changes during keratinization. *Anat. Rec.* **156**, 269–282.
7. Filshie, B. K., and Rogers, G. E. (1961). The fine structure of α-keratin. *J. Mol. Biol.* **3**, 784–786.
8. Forslind, B., and Swanbeck, G. (1966). Keratin formation in the hair follicle. I. An ultrastructural investigation. *Exp. Cell Res.* **43**, 191–209.
9. Fraser, R. D. B. (1969). Keratins. *Sci. Amer.* **221** (2), 87–96.
10. Hay, E. D. and Revel, J. P. (1969). Fine structure of the developing avian cornea. *In* "Monographs in Developmental Biology," Vol. I. S. Karger, New York. 144 pp.
11. Lavker, R. M. and Matoltsy, A. G. (1970). Formation of horny cells. The fate of cell organelles and differentiation in ruminal epithelium. *J. Cell Biol.* **44**, 501–512.
12. Lundgren, H. P., and Ward, W. (1963). The keratins. *In* "Ultrastructure of Protein Fibers" (R. Borasky, ed.), pp. 39–122. Academic Press, New York.
13. Lyne, A. G., and Short, B. F. (1965). "Biology of the Skin and Hair Growth." Angus and Robertson, Sydney.
14. Matoltsy, A. G. (1962). Structural and chemical properties of keratin-forming tissues. *In* "Comparative Biochemistry" (M. Florkin and H. S. Mason, eds.), Vol. IV, pp. 343–369. Academic Press, New York.
15. Matoltsy, A. G. (1965). Soluble prekeratin. *In* "Biology of the Skin and Hair Growth." (A. G. Lyne and B. F. Short, eds.), pp. 291–305. Angus and Robertson, Sydney.
16. Matoltsy, A. G., and Parakkal, P. F. (1967). Keratinization. *In* "Ultrastructure of Normal and Abnormal Skin" (A. Zelickson, ed.), pp. 76–104. Lea & Febiger, Philadelphia, Pennsylvania.
17. Mercer, E. H. (1961). "Keratin and Keratinization." Pergamon, New York.
18. Millward, G. R. (1970). The substructure of α-keratin microfibrils. *J. Ultrastruct. Res.* **31**, 349–355.
19. Montagna, W. (1962). "The Structure and Function of Skin." Academic Press, New York.
20. Montagna, W., and Lobitz, W., eds. (1964). "The Epidermis." Academic Press, New York.
21. Pauling, L., and Corey, R. B. (1953). Compound helical configurations of polypeptide chains: structure of proteins of the α keratin type. *Nature (London)* **171**, 59–61.
22. Rogers, G. E. (1964). Structural and biochemical features of the hair follicle. *In* "The Epidermis" (W. Montagna and W. Lobitz, eds.), pp. 179–232. Academic Press, New York.
23. Rudall, K. M. (1947). X-ray studies of the distribution of protein chain types in the vertebrate epidermis. *Biochim. Biophys. Acta* **1**, 549–562.
24. Rudall, K. M. (1952). The proteins of the mammalian epidermis. *Advan. Protein Chem.* **7**, 253–290.
25. Spearman, R. I. C. (1966). The keratinization of epidermal scales, feathers and hairs. *Biol. Rev.* **41**, 59–96.
26. Weinstock, M., and Wilgram, G. F. (1970). Fine structural observations on the formation and enzymatic activity of keratinosomes in mouse tongue filiform papillae. *J. Ultrastruct. Res.* **30**, 262–274.
27. Wessells, N. K. (1967). Differentiation of epidermis and epidermal derivatives, *N. Eng. J Med.* **277**, 21–33.
28. Wolff, K., and Holubar, K. (1967). Odland-Körper (Membrane Coating Granules, Keratinosomen) als epidermal Lysosomen. *Arch. Klin. Exp. Dermatol.* **231**, 1–19.

SELECTED BIBLIOGRAPHY

Fish

Bierther, M. (1970). Die Chloridzellen des Stichlings. *Z. Zellforsch.* **107**, 421–446.
Brown, G. A., and Wellings, S. R. (1970). Electron microscopy of the skin of the teleost *Hippoglossoides elassodon*. *Z. Zellforsch.* **103**, 149–169.
Henrikson, R. C., and Matoltsy, A. G. (1968). The fine structure of teleost epidermis. I. Introduction and filament containing cells. *J. Ultrastruct. Res.* **21**, 194–212. II. Mucous cells. *J. Ultrastruct. Res.* **21**, 213–221. III. Club cells and other cell types. *J. Ultrastruct. Res.* **21**, 222–232.
Shirai, N., and Utida, S. (1970). Development and degeneration of the chloride cell during seawater and freshwater adaptation of the Japanese eel, *Anguilla japonica*. *Z. Zellforsch.* **103**, 247–264.
Wellings, S. R., Chuinard, R. G., and Cooper, R. A. (1967). Ultrastructural studies of normal skin and epidermal papillomas of the flathead sole *Hippoglossoides elassodon*. *Z. Zellforsch.* **78**, 370–387.

Amphibia

Chapman, G. B., and Dawson, A. B. (1961). Fine structure of the larval anuran epidermis, with special reference to the figures of Eberth. *J. Biophys. Biochem. Cytol.* **10**, 425–435.
Farquhar, M. G., and Palade, G. E., (1965). Cell junctions in amphibian skin. *J. Cell Biol.* **26**, 263–291.

Kelly, D. E. (1966). The Leydig cells in larval amphibian epidermis. Fine structure and function. *Anat. Rec.* **154**, 685–700.

Parakkal, P. F., and Matoltsy, A. G. (1954). A study of the fine structure of the epidermis of *Rana pipiens*. *J. Cell Biol.* **20**, 85–94.

Pillai, P. A. (1962). Electron microscopic studies on the epidermis of newt with an inquiry into the problem of induced neoplasia. *Protoplasma* **55**, 10–62.

Voute, C. L. (1963). An electron microscopic study of the skin of the frog (*Rana pipiens*). *J. Ultrastruct. Res.* **9**, 497–510.

Reptiles

Alexander, N. J. (1970). Comparison of α and β keratin in reptiles. *Z. Zellforsch.* **110**, 153–165.

Alexander, N. J., and Parakkal, P. F. (1969). Formation of α- and β-type keratin in lizard epidermis during the molting cycle. *Z. Zellforsch.* **101**, 72–87.

Bryant, S. V., Breathnach, A. S., and Bellaris, A. A. (1967). Ultrastructure of the epidermis of the lizard (*Lacerta vivipara*) at the resting stage of the sloughing cycle. *J. Zool.* **152**, 209–219.

Ernst, V., and Ruibal, R. (1966). The structure and development of the digital lamellae of lizards. *J. Morphol.* **120**, 233–266.

Flaxman, B. A., Maderson, P. F. A., Szabo, G., and Roth, S. I. (1968). Control of cell differentiation in lizard epidermis *in vitro*. *Develop. Biol.* **18**, 354–374.

Horstmann, E. (1964). Elektronenmikroskopische Untersuchungen an der Epidermis von Reptilien. *Anat. Anz.* **113**, 87–93.

Lillywhite, H. B., and Maderson, P. F. A. (1968). Histological changes in the epidermis of the subdigital lamellae of *Anolis carolinensis* during the shedding cycle. *J. Morphol.* **125**, 379–402.

Maderson, P. F. A., and Licht, P. (1967). Epidermal morphology and sloughing frequency in normal and prolactin treated *Anolis carolinensis* (Iguanidae, Lacertilia). *J. Morphol.* **123**, 157–172.

Roth, S. I., and Jones, W. A. (1967). The ultrastructure and enzymatic activity of the boa constrictor (*Constrictor constrictor*) skin during the resting phase. *J. Ultrastruct. Res.* **18**, 304–323.

Roth, S. I., and Jones, W. A. (1970). The ultrastructure of epidermal maturation in the skin of the boa constrictor (*Constrictor constrictor*). *J. Ultrastruct. Res.* **32**, 69–93.

Rudall, K. M. (1947). X-ray studies of the distribution of protein chain types in the vertebrate epidermis. *Biochim. Biophys. Acta* **1**, 549–562.

Birds

Bell, E., and Thathachari, Y. R. (1963). Development of feather keratin during embryogenesis of the chick. *J. Cell Biol.* **16**, 215–223.

Filshie, B. K., and Rogers, G. E. (1961). An electron microscope study of the fine structure of feather keratin. *J. Cell Biol.* **13**, 1–12.

Matoltsy, A. G. (1970). Keratinization of the avian epidermis. An ultrastructural study of the newborn chick skin. *J. Ultrastruct. Res.* **29**, 438–458.

Mottet, N. K., and Jensen, H. M. (1968). The differentiation of chick embryonic skin. An electron microscopic study with a description of a peculiar epidermal cytoplasmic ultrastructure. *Exp. Cell Res.* **52**, 261–283.

Parakkal, P. F., and Matoltsy, A. G. (1968). An electron microscopic study of developing chick skin. *J. Ultrastruct. Res.* **23**, 403–416.

Rawles, M. E. (1965). Tissue interactions in the morphogenesis of the feather. *In* "Biology of Skin and Hair Growth" (A. G. Lyne, and B. F. Short, eds.), pp. 105–128. Angus and Robertson, Sydney.

Sengel, P., and Rasaonen, M. (1969). Modifications ultrastructurales au cours de l'histogenèse de la peau chez l'embryon de poulet. *Arch. Anat. Microsc. Morphol. Exp.* **58**, 77–96.

Mammals

Breathnach, A. S., and Wylie, L. M. (1965). Fine structure of cells forming the surface layer of the epidermis in human fetuses at fourteen and twelve weeks. *J. Invest. Dermatol.* **45**, 179.

Bonneville, M. A. (1968). Observations on epidermal differentiation in the fetal rat. *Amer. J. Anat.* **123**, 147–164.

Brody, I. (1968). The Epidermis. *In* "Jadassohn's Handbuch der Haut-und Geschlechtskranheiten. Ergänzungswerk. Normale und pathologische Anatomie der Haut I" (O. Gans and G. K. Steigleder, eds.), I/1, p. 1. Springer, New York.

Cooper, R. A., Cardiff, R. D., and Wellings, S. R. (1967). Ultrastructure of vaginal keratinization in estrogen treated immature Balb/c mice. *Z. Zellforsch.* **77**, 377–403.

Eddy, E. M., and Walker, B. E. (1969). Cytoplasmic fine structure during hormonally controlled differentiation in vaginal epithelium. *Anat. Rec.* **164**, 205–218.

Farbman, A. I. (1964). Electron microscope study of a small cytoplasmic structure in rat oral epithelium. *J. Cell Biol.* **21**, 491–495.

Jensen, H. (1970). Two types of keratohyalin granules. *J. Ultrastr. Res.* **33**, 95–115.

Matoltsy, A. G., and Parakkal, P. F. (1965). Membrane-coating granules of keratinizing epithelia. *J. Cell Biol.* **24**, 297–307.

Odland, G. (1964). Tonofilaments and keratohyalin. *In* "The Epidermis" (W. Montagna and W. Lobitz, eds.). Academic Press, New York.

Weinstock, M., and Wilgram, G. F. (1970). Fine structural observations on the formation and enzymatic activities of keratinosomes in mouse tongue filiform papillae. *J. Ultrastruct. Res.* **30**, 262–274.

Hair and Fingernail

Birbeck, M. S. C., and Mercer, E. H. (1957). The electron microscopy of the human hair follicle. I. Introduction and hair cortex. *J. Biophys. Biochem. Cytol.* **3**, 203–214. II. The hair cuticle. *J. Biophys. Biochem. Cytol.* **3**, 215–222. III. The inner root sheath and trichohyalin. *J. Biophys. Biochem. Cytol.* **3**, 223–230.

Chase, H. B. (1964). Growth of the hair. *Physiol. Rev.* **34**, 113–126.

Hasimoto, K., Gross, B. G., Nelson, R., and Lever, W. F. (1966). The ultrastructure of the skin of human embryos. III. The formation of the nail in 16–18 week old embryos. *J. Invest. Dermatol.* **47**, 205–217.

Parakkal, P. F. (1969). The fine structure of anagen hair follicle in the mouse. *In* "Hair Growth" (W. Montagna and R. Dobson, eds.), pp. 441–469. Pergamon, New York.

Parakkal, P. F. (1969). Ultrastructural changes of the basal lamina during the hair growth cycle. *J. Cell Biol.* **40**, 561–564.

Parakkal, P. F. (1969). Role of macrophages in collagen resorption during hair growth cycle. *J. Ultrastruct. Res.* **29**, 201–217.

Parakkal, P. F. (1970). Morphogenesis of the hair follicle during catagen. *Z. Zellforsch.* **107**, 174–186.

Puccinelli, V. A., Caputo, R., and Ceccarelli, B. (1967). The structure of human hair follicle and hair shaft: an electron microscope study. *G. Ital. Dermatol.* **108**, 1–46.

Roth, S. I. (1965). The cytology of the murine resting (telogen) hair follicle. *In* "Biology of Skin and Hair Growth" (A. G. Lyne and B. F. Short, eds.), pp. 233–250. Angus and Robertson, Sydney.

Roth, S. I. (1967). Hair and nail. *In* "Ultrastructure of Normal and Abnormal Skin" (A. S. Zelickson, ed.), pp. 105–131. Lea & Febiger, Philadelphia, Pennsylvania.

Rogers, G. E. (1964). Structure and biochemical features of the hair follicles. *In* "The Epidermis" (W. Montagna and W. Lobitz, eds.), pp. 179–236. Academic Press, New York.

Thorndike, E. E. (1968). A microscopic study of the marmoset claw and nail. *Amer. J. Phys. Anthropol.* **28**, 247–262.

Zaias, N., and Alvarez, J. (1968). The formation of the primate nail plate. An autoradiographic study in squirrel monkey. *J. Invest. Dermatol.* **51**, 121–136.

Printed in Dunstable, United Kingdom

85221344R00038